Interactive Beginning Algebra

Interactive Mathematics

Skip Forward

Print

Quit-Bookmark

Review

E-mail to Mentor

Subject: Chapter 1, Unit 5, Frame 1

Mentor1on1

Send Now Quote Attach Address Stop

Subject: Re: Chapter 1, Unit 5, Frame 1

Addressing

Mail To: Mentor1on1

Cc:

Hey Mentor,

Thanks for the help!

Robert H. Alwin
Robert D. Hackworth

H&H Interactive Mathematics Series

H&H Publishing Company, Inc.
Clearwater, Florida

H&H Publishing Company, Inc.

1231 Kapp Drive
Clearwater, FL 33765
(727) 442-7760
(800) 366-4079
FAX (727) 442-2195
email: Interactive@HHPublishing.com
web site: www.HHPublishing.com

INTERACTIVE BEGINNING ALGEBRA

Robert H. Alwin
Robert D. Hackworth

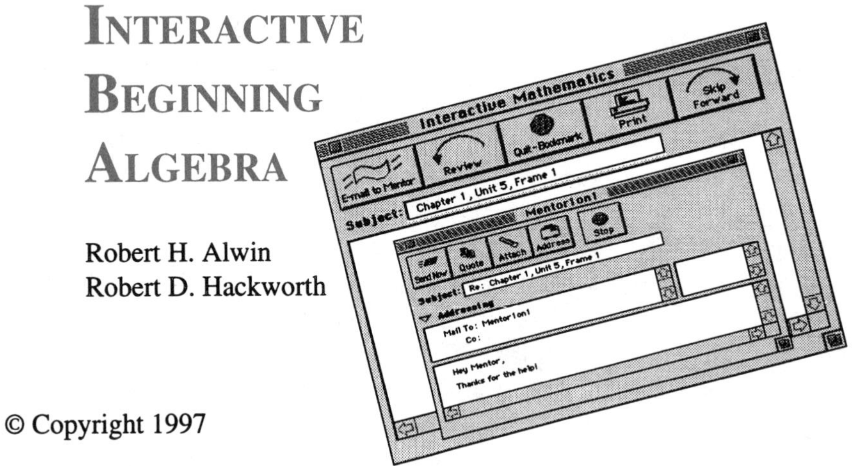

ISBN 0-943202-63-9

Library of Congress Catalog Number 97-072104

This text in previous versions was originally published by Prentice-Hall, Inc.

Printing is the lowest number: 10 9 8 7 6 5 4 3

PREFACE

Interactive Beginning Algebra is designed to aid the student in learning its material.

The text possesses strong teaching qualities that virtually assure the conscientious student of steady, confident progress. At the same time, the text's interactive processes assure the student of continuous opportunities to get help where and when it is needed. Solid mathematics, quality instruction, and personal assistance are the hallmark of this series of texts.

The text is designed in a format that constantly creates an interaction between the student, the mathematics to be learned, and an experienced teacher either in a classroom or in cyberspace. The majority of the book is a series of questions/answers that achieves a true dialogue with its student reader. Most of the time, the dialogue teaches the mathematics effectively, and students learn easily without benefit of outside assistance. Sometimes, however, the dialogue may be insufficient; it is in such cases where the student can find extra, personal assistance through conversation with a friend or teacher. These helpful conversations can occur in person, but also are available with a cyberspace instructor who is always as close as the nearest computer.

The quality of the mathematical content and instructional use of this text have been thoroughly documented in a wide range of situations and with great numbers of students. Specifically, there are no units or chapters where common difficulties can be expected. When a student has a problem understanding the material, the difficulties will be unique to that student. Something of importance was overlooked, misunderstood, etc. That is the reason for providing personal assistance in those instances and getting the student back into the flow of the book's dialogue. The personal assistance will both answer the questions of individual students and give guidance in the best ways to study mathematics.

Because of the design of this book, the student can expect other dramatic improvements in the study and learning of mathematics. One of those advantages is the student's ability to vary the pace of instruction to match their ability to learn. At times, a student needs to work slowly through the instruction and carefully ponder each new concept. At other times, the student grasps the material quickly and can accelerate their learning rapidly through the material. As a result, most students will move through this material more quickly than in a regular class situation.

Another great advantage of this book is the fact that students have an opportunity to learn to ask intelligent, specific questions about the mathematics in a context that encourages and does not threaten the student. Study skill deficiencies can be corrected within the context of these materials while maintaining traditional levels of achievement and rigor.

The content of this book is comparable to a first semester of Algebra as it is normally taught in high school. Chapters 1-3 deal only with the counting numbers so that the first concepts of Algebra can be introduced in a simple, already known context. Chapter 1 teaches the generalizations of arithmetic that need to be understood. In Chapter 2, the first skills of algebra, including the simplification of open expressions, are taught. Chapter 3 teaches the basic skills and concepts of solving equations. Chapters 4-6 repeat the concepts of Chapters 1-3, but apply this learning to the integers. Then Chapters 7-9 repeat the cycle for the rational numbers. The spiraling effect on learning is obvious in the treatment of the counting numbers, integers, and rational numbers. Each chapter ends with an Application Unit teaching word problem-solving processes. Unlike most text presentations of word problems, the student is taught a process for breaking down the information of a word problem and developing methods for extending these processes to more difficult problems to be encountered later.

The book contains many opportunities for the student to make wise decisions about the level of understanding achieved. Each chapter begins with an Objectives Test that illustrates, by example, all of the concepts and skills which are to be presented. Each problem is accompanied by a designation showing where it will be taught in the chapter. Each chapter is broken into units in a way that divides the content into topics of such length that it gives the student a way of organizing their own studying to make it most effective. Most units are designed to be completed in a single study session and, for best results, should be handled in that way. The unit ends with a Feedback which is an opportunity for the student to test the level of understanding achieved. It is strongly suggested that the Feedback be taken at the beginning of the next study session so that some time intervenes between the learning and assessment. The chapter ends with a Mastery Test that is similar to the Objectives Test. When a student can do all the problems correctly on a Mastery Test, it indicates a complete understanding of all the material in the chapter. Again, each problem is accompanied by a designation showing where it was taught in the chapter. Answers for all problems in the Tests and Feedbacks are included at the back of the book. Good learning can occur only when the student is aware of what is known and what still needs to be learned.

We are greatly indebted to thousands of teachers and hundreds of thousands of students who have successfully used our previous materials and given us the benefit of their own experiences and ideas. The strength of this new book is a result of that special kind of study of the work and challenges we face in teaching mathematics to students — some in desperate need for overcoming past experiences that were markedly unsuccessful and have interfered with motivations to try again.

Robert H. Alwin
Robert D. Hackworth

DIRECTIONS TO THE STUDENT

It is important to use this text correctly to achieve its full benefits. Its format encourages you to become a responsible participant in your own learning as you enter a dialogue with the book. The majority of the book involves giving you information, asking you a question about it, waiting for your answer, and then confirming or denying your understanding. You need to respond thoughtfully just as if you were in a conversation with a close friend about some matter of particular interest.

Each of these question-answer cycles is numbered and is called a frame. The first sentence of a frame will generally tell you what to look for or focus upon. Then you will be asked a question about the focus of the frame. Think about it and then write your answer. After you have written your answer, look in the right-hand column to see whether you completely understood the purpose of the frame. When you have completely understood, proceed to the next frame. When something is not completely understood, try to see what was not clearly communicated. If the misunderstanding continues, then seek personal assistance and resolve the matter before continuing.

Each frame contains something to be learned. Read the first sentence to find where the authors want you to focus your attention. Writing your answers to the question is important because it requires a commitment and that commitment is necessary if you are to learn. Checking your answer immediately is important. When the answer is correct, the learning is reinforced and, therefore, remembered. When the answer in incorrect, you are immediately alerted to some misunderstanding. This is not necessarily bad because many of your most important learning experiences will occur following some awareness of a problem. Treat these situations as good learning opportunities.

Each chapter begins with a set of problems that illustrate the objectives of the chapter. You are not expected to know how to do the problems on an Objectives Test. Each problem indicates the unit in which it will be learned. Use these problems to acquaint yourself with what is to be learned in the chapter and do not allow yourself to be frightened by any new symbols or seemingly difficult situations. In the rare case that all the problems on an Objectives Test are truly easy, it is possible to consider jumping to the Chapter Mastery Test.

Each unit concludes with a Feedback exercise for you to test your understanding of the particular skills learned in the unit. In general, it is best not to take the Feedback immediately after finishing the frames of a unit. Give yourself some time before assessing your understanding; it will improve your progress.

Each chapter ends with a Mastery Test for you to assess your understanding of all the material in the chapter. Each problem is accompanied by a designation indicating where it is taught in the chapter. Use that information to restudy any portions of the chapter that the text indicates you have not mastered.

Throughout your study of this book, put your emphasis on understanding the material rather than memorizing without understanding. Regardless of your past experiences with learning mathematics, do not be tempted into memorizing. Those that memorize may remember for a week or a month, but then suffer a complete loss and must start anew. Those who understand, rather than memorize, have something that will sustain them throughout life. You'll find that every problem taught in this book has a purpose which is explained to you. Memorizing isn't necessary to your successful dialogue. Understanding will definitely improve your achievement level and your satisfaction.

CONTENTS

PREFACE III
DIRECTIONS TO THE STUDENT V

CHAPTER 1 OBJECTIVES XII
CHAPTER 1 THE ARITHMETIC OF THE COUNTING NUMBERS 1
UNIT 1: THE SET OF COUNTING NUMBERS 1
UNIT 2: EVALUATING ADDITION EXPRESSIONS 8
UNIT 3: EVALUATING SUBTRACTION EXPRESSIONS 18
UNIT 4: EVALUATING MULTIPLICATION EXPRESSIONS 26
UNIT 5: EVALUATING DIVISION EXPRESSIONS 34
UNIT 6: EVALUATING NUMERICAL EXPRESSIONS INVOLVING MORE THAN ONE OPERATION 40
UNIT 7: FACTORS AND MULTIPLES OF COUNTING NUMBERS 50
UNIT 8: APPLICATIONS 58
SUMMARY 62
MASTERY TEST 63

CHAPTER 2 OBJECTIVES 64
CHAPTER 2 THE ALGEBRA OF THE COUNTING NUMBERS 65
UNIT 1: NUMERICAL AND OPEN EXPRESSIONS 65
UNIT 2: EVALUATING NUMERICAL EXPRESSIONS 70
UNIT 3: COMMUTATIVE AND ASSOCIATIVE LAWS OF ADDITION 77
UNIT 4: COMMUTATIVE AND ASSOCIATIVE LAWS OF MULTIPLICATION AND THE MULTIPLICATION LAW OF ONE 86
UNIT 5: THE DISTRIBUTIVE LAW AND SIMPLIFYING OPEN EXPRESSIONS WITH LIKE TERMS 91
UNIT 6: SIMPLIFYING OPEN EXPRESSIONS 97
UNIT 7: APPLICATIONS 102
SUMMARY 106
MASTERY TEST 107

CHAPTER 3 OBJECTIVES 108

CHAPTER 3 SOLVING EQUATIONS WITH THE COUNTING NUMBERS 109

UNIT 1: STATEMENTS AND OPEN SENTENCES 109
UNIT 2: SOLVING EQUATIONS 114
UNIT 3: FINDING TRUTH SETS FOR EQUATIONS USING THE SET OF COUNTING NUMBERS 116
UNIT 4: SOLVING EQUATIONS AND CHECKING THE SOLUTIONS 122
UNIT 5: APPLICATIONS 126
SUMMARY 130
MASTERY TEST 131

CHAPTER 4 OBJECTIVES 132

CHAPTER 4 THE ARITHMETIC OF THE INTEGERS 133

UNIT 1: THE SET OF INTEGERS 133
UNIT 2: THE NUMBER LINE FOR THE SET OF INTEGERS 139
UNIT 3: MULTIPLICATION OF INTEGERS 151
UNIT 4: SUBTRACTION OF INTEGERS AND EVALUATING NUMERICAL EXPRESSIONS 158
UNIT 5: EVALUATING NUMERICAL EXPRESSIONS INVOLVING MULTIPLICATION AND EXPONENTS 167
UNIT 6: PRIME AND COMPOSITE NUMBERS 173
UNIT 7: APPLICATIONS 178
SUMMARY 182
MASTERY TEST 183

CHAPTER 5 OBJECTIVES 184

CHAPTER 5 THE ALGEBRA OF THE INTEGERS 185

UNIT 1: EVALUATING NUMERICAL EXPRESSIONS 185
UNIT 2: SIMPLIFYING ADDITION EXPRESSIONS 189
UNIT 3: SIMPLIFYING MULTIPLICATION EXPRESSIONS 194
UNIT 4: SIMPLIFYING OPEN EXPRESSIONS 197
UNIT 5: REMOVING PARENTHESES TO SIMPLIFY OPEN EXPRESSIONS 204
UNIT 6: SIMPLIFYING POWER EXPRESSIONS 208
UNIT 7: APPLICATIONS 214
SUMMARY 218
MASTERY TEST 219

Chapter 6 Objectives 220

Chapter 6 Solving Equations with the Integers 221

Unit 1: Solving Simple Equations 221
Unit 2: Generating Equivalent Equations 226
Unit 3: Finding Solutions 231
Unit 4: Checking Solutions 238
Unit 5: Applications 241
Summary 244
Mastery Test 245

Chapter 7 Objectives 246

Chapter 7 The Arithmetic of the Rational Numbers 247

Unit 1: Defining Rational Numbers 247
Unit 2: Equal Rational Numbers 254
Unit 3: Simplifying Rational Numbers 261
Unit 4: Multiplication of Rational Numbers 266
Unit 5: Writing Pairs of Rational Numbers with a Common Denominator 275
Unit 6: Adding Rational Numbers 282
Unit 7: Reciprocals and Division of Rational Numbers 289
Unit 8: Applications 298
Summary 301
Mastery Test 302

Chapter 8 Objectives 304

Chapter 8 The Algebra of the Rational Numbers 305

Unit 1: Simplifying Addition Expressions 305
Unit 2: Simplifying Multiplication Expressions 308
Unit 3: Simplifying Open Expressions 312
Unit 4: Multiplication Using Exponents 320
Unit 5: Division Using Exponents 330
Unit 6: Applications 336
Summary 339
Mastery Test 340

CHAPTER 9 OBJECTIVES 342

CHAPTER 9 SOLVING EQUATIONS WITH THE RATIONAL NUMBERS 343

UNIT 1: FINDING SOLUTIONS BY ADDING OPPOSITES 343
UNIT 2: FINDING SOLUTIONS BY MULTIPLYING BY RECIPROCALS 348
UNIT 3: FINDING TRUTH SETS 356
UNIT 4: SOLVING AND CHECKING FRACTIONAL EQUATIONS 361
UNIT 5: APPLICATIONS 369
SUMMARY 372
MASTERY TEST 373

ANSWERS FOR ALL TESTS AND FEEDBACK EXERCISES 375

INDEX 387

Chapter 1 Objectives

The following problems illustrate the objectives of this chapter. At this time you are not expected to know how to do these problems. However, if all of these problems are thoroughly understood, proceed directly to the Chapter 1 Mastery Test. The number in parentheses which follows each problem indicates the unit in which it can be learned.

1. Write the set of even numbers between 11 and 19. (1)

2. Write the set of odd numbers between 17 and 24. (1)

3. Write the set of odd numbers between 11 and 13. (1)

4. Is the statement $14 \notin \{1, 2, 3, \ldots\}$ true or false? (1)

5. Is the statement $17 \in \{3, 4, 5, 17, 19\}$ true or false? (1)

6. Is the statement $13 \notin \{\ \}$ true or false? (1)

7. Evaluate. $(7 + 3) + 6$ (2)

8. Evaluate. $26 - (28 - 17)$ (3)

9. Evaluate. $6 \bullet (3 \bullet 4)$ (4)

10. Evaluate. $35 \div (15 \div 3)$ (5)

11. Evaluate. $2 + 4 \bullet 8$ (6)

12. Evaluate. $9 \bullet 3 + 5$ (6)

13. Evaluate. $3 \bullet [4 + (3 + 2)]$ (6)

14. Evaluate. $[2 \bullet (1 + 5)] + 7$ (6)

15. Write the set of all counting number factors of 15. (7)

16. Find the highest common factor of 16 and 40. (7)

17. Find the least common multiple of 6 and 15. (7)

CHAPTER 1
THE ARITHMETIC OF THE COUNTING NUMBERS

UNIT 1: THE SET OF COUNTING NUMBERS

The following mathematical terms are crucial to an understanding of this unit.

Set	Even numbers
Odd numbers	Between
Empty set	Element
Member	Counting numbers

1
In mathematics a collection of numbers is called a **set**.
For a golfer, the group of clubs is called a set of clubs.
In mathematics, the collection of numbers 3, 5, and 7 is
called the ______ of numbers.

set

2
A group of encyclopedias is a **set** of encyclopedias.
Similarly, the collection of numbers 2, 4, 6, and 8 is
called a ______ of numbers.

set

3

The collection consisting of 5, 9, 11, and 15 is called a ______ of numbers.

set

4

Braces, { }, are used to indicate a set. The set of numbers 2, 5, and 9 is shown as {2, 5, 9}. Using braces, show the set of numbers 8, 9, and 34.

{8, 9, 34}

5

Using braces, write the set of numbers 9 and 5.

{5, 9} or {9, 5}
(Either order is correct.)

6

The collection of all whole numbers greater than 5 and less than 8 is shown by the set {6, 7}. Use braces to show the set of whole numbers greater than 19 and less than 23.

{20, 21, 22}

7

Write the set of whole numbers greater than 7 and less than 11.

{8, 9, 10}

8

The set of **even numbers** greater than 1 and less than 9 is {2, 4, 6, 8}. Write the set of even numbers greater than 1 and less than 7.

{2, 4, 6}

9

{6} is the set of even numbers greater than 5 and less than 8. Write the set of even numbers greater than 16 and less than 19.

{18}

10

The set of whole numbers greater than 4 and less than 7 can be stated as the set of whole numbers **between** 4 and 7. {8, 9} is the set of whole numbers greater than 7 and less than 10, or the set of whole numbers ______ 7 and 10.

between

11

Write the set of even numbers between 88 and 94.

{90, 92}

12

The **empty set**,{ }, is the set which has no objects or numbers in it. How many elements are in the empty set,{ }?

None

13

{ } would show the set of even numbers between 10 and 20 that are greater than 25. Write the set of odd numbers between 20 and 30 that are less than 16.

{ }

14

Write the set of whole numbers between 25 and 30 that are greater than 100.

{ }

15

{1, 3, 5, 7, 9} is the set of **odd numbers** less than 11. Write the set of odd numbers less than 7.

{1, 3, 5}

16

{5, 7, 9, 11, 13} is the set of odd numbers greater than 4 and less than 15. Write the set of odd numbers greater than 7 and less than 12.

{9, 11}

17

{ } indicates the empty set. Write the set of whole numbers between 4 and 5.

{ }

18

4 is in the set {4, 7, 9, 15, 20}. 7 and 9 are in the set. 8 is not in the set. Is 15 in the set?

Yes

19

5 is in the set {2, 5, 8, 11}. Is 11 in the set {2, 5, 8, 11}?

Yes

20

6 is not in the set {3, 4, 5, 9}. Is 5 in the set {2, 4, 6}?

No

21

27 is not in the empty set. Is 71 in the empty set { }?

No

22

7 is a **member** or **element** of the set $\{1, 3, 7, 10\}$, since 7 is in the set. 10 is also a member or ______ of the set $\{1, 3, 7, 10\}$.

element

23

Is 7 an element of $\{3, 6, 7, 9\}$?

Yes

24

47 is an ______ of $\{102, 19, 47, 316, 32\}$.

element or member

25

"$\in$" is a mathematical symbol which means "is an element (or member)" of a set. Since 3 is an element of the set $\{1, 2, 3, 5\}$, it is true to write $3 \in \{1, 2, 3, 5\}$. Use "$\in$" to write 5 is an element of $\{1, 2, 3, 5\}$.

$5 \in \{1, 2, 3, 5\}$

26

Use the symbol for element "$\in$" to write "93 is an element of the set $\{13, 93, 49\}$."

$93 \in \{13, 93, 49\}$

27

$5 \in \{2, 5, 8, 9\}$. True or false?

True

28

$7 \in \{2, 4, 6, 8\}$. True or false?

False

29

"$\notin$" is a mathematical symbol for "is not an element of a set." 7 is not an element of $\{2, 4, 6, 8\}$ is written $7 \notin \{2, 4, 6, 8\}$. Using the symbol "$\notin$" show that 5 is not an element of $\{2, 4, 6, 8\}$.

$5 \notin \{2, 4, 6, 8\}$

30

Use the symbol "$\notin$" to indicate that 15 is not an element of $\{4, 8, 12, 19\}$.

$15 \notin \{4, 8, 12, 19\}$

31

Use the correct symbol, "$\in$" or "$\notin$" to fill in the blank. 5 ___ $\{2, 4, 8\}$ is a true statement.

$\notin$

32
Fill in the blank. 17 ______ {5, 10, 15, 20} is a true statement.

$\notin$

33
In **counting** the stars in the box, ☆ ☆ ☆ ☆ , the numbers 1, 2, 3, 4 are used. What numbers are used in counting the number of stars in this box? ☆ ☆ ☆

1, 2, 3

34
What numbers are used in counting the triangles in this box? Δ Δ Δ Δ Δ Δ Δ

1, 2, 3, 4, 5, 6, 7

35
What numbers are used to count the stars in this box?
☆ ☆ ☆ ☆ ☆ ☆ ☆ ☆ ☆ ☆ ☆ ☆

1, 2, 3, 4, 5, 6, 7, 8, 9, 10, 11, 12

36
Write the set to show the numbers used to count the following stars:
☆ ☆ ☆ ☆ ☆ ☆ ☆ ☆ ☆ ☆ ☆ ☆ ☆ ☆ ☆

{1, 2, 3, 4, 5, 6, 7, 8, 9, 10, 11, 12, 13, 14, 15}

37
Is it possible to show the set of numbers used in counting 50 stars?

Yes

38
Is it possible to show the set, by listing the elements in braces, of numbers used in counting 938 objects?

Yes

Instead of listing the numbers from 1 to 938 to show the elements of the set, mathematicians use three dots (. . .) in the following manner to show the set:

{1, 2, 3, . . . , 938}

[Note: The three dots mean "and so forth" or to continue in the same pattern and to stop at the last number in the set.]

39

The set of numbers used to count 25 stars is shown, using the three-dot symbol, as {1, 2, 3, . . . , 25}. Write the set of numbers used to count 57 stars.

{1, 2, 3, . . . , 57}

40

Write the set to show the numbers used to count 291 stars.

{1, 2, 3, . . . , 291}

41

48 is an element of the set {1, 2, 3, . . . , 457} because 48 is between 1 and 457 and the three dots indicate that the counting is to start at 1 and continue consecutively to 457. Is 196 in the set?

Yes

42

Is 487 in the set {1, 2, 3, . . . , 968}?

Yes

43

The largest number in the set {1, 2, 3, . . . , 47} is 47. What is the largest number in the set {1, 2, 3, . . . , 32}?

32

44

The smallest number in {1, 2, 3, . . . , 584} is 1. What is the smallest number in the set {1, 2, 3, . . . , 324}?

1

45

What is the smallest number in the set {1, 2, 3, . . . , 74}?

1

46

If the set ends with three dots instead of a number, the counting continues without stopping. The **set of all counting numbers** is {1, 2, 3, . . . }. Is there a largest number in the set {1, 2, 3, . . . }?

No

47

Is 1 the smallest counting number in the set {1, 2, 3, . . . , 578}?

Yes

48

Is there a largest counting number in the set {1, 2, 3, . . . }?

No

49
Every number used in counting is in the set {1, 2, 3, . . . }.
Is 3,469 in the set {1, 2, 3, . . . }?

Yes

50
The set {1, 2, 3, . . . } is called the **set of counting numbers**.
Is 7 in the set of counting numbers?

Yes

51
Write the set of counting numbers.

{1, 2, 3, . . . }

FEEDBACK UNIT 1

This quiz reviews the preceding unit. Answers are at the back of the book.

1. In mathematics a collection of numbers is called a ______.
2. Using braces, write the set of numbers 2, 4, 8, and 9.
3. Write the set of whole numbers greater than 7 and less than 10.
4. Write the set of whole numbers greater than 10 and less than 7.
5. Write the set of even numbers greater than 7 and less than 12.
6. Write the set of odd numbers between 5 and 12.
7. Write the set of odd numbers between 21 and 23.
8. Is 14 in the set {2, 9, 13, 15}?
9. Is 6 in the set {1, 3, 5, 6, 10}?
10. Is 23 in the set { }?
11. Use the symbols "∈" or "∉" to make each of the following a true statement.
    a. 9 ______ {1, 3, 5, 7, 9, 13}.
    b. 13 ______ {2, 6, 8, 10, 11, 15}.
    c. 96 ______ {13, 47, 83, 95}.
    d. 47 ______ {2, 41, 42, 44, 47, 51, 96}.
    e. 27 ______ { }.
12. 47 ∈ {1, 2, 3, . . . , 49}. True or false?
13. 93 ∉ {1, 2, 3, . . . }. True or false?
14. 2 ∈ {1, 2, 5, 8}. True or false?
15. 32 ∈ { }. True or false?
16. 23 ∈ {1, 2, 3, . . . , 20}. True or false?
17. 17 ∈ {1, 3, 16, 19}. True or false?
18. The set of counting numbers is {1, 2, 3, . . . }. True or false?
19. 16 ∉ { }. True or false?

UNIT 2: EVALUATING ADDITION EXPRESSIONS

The following mathematical terms are crucial to an understanding of this unit.

Addition	Expression
Sum	Evaluate the expression
Parentheses	Square brackets

1
Count the number of stars in each of the boxes.
☆ ☆ ☆ | ☆ ☆ ☆ ☆ ☆

3, 5

2
If the stars from both boxes in the previous frame are put into one box, how many stars are there?
☆ ☆ ☆ ☆ ☆ ☆ ☆ ☆

8

3
Count the stars in each box.
☆ ☆ ☆ ☆ ☆ ☆ | ☆ ☆ ☆ ☆

6, 4

4
Count the stars in both boxes in the previous frame.
☆ ☆ ☆ ☆ ☆ ☆ ☆ ☆ ☆ ☆

10

5
Count the stars in each box.
☆ ☆ ☆ ☆ ☆ ☆ ☆ ☆ | ☆ ☆ ☆ ☆ ☆

8, 5

6
Count all the stars in both boxes in the previous frame.

13

7
By placing all the stars in one box and counting the number of stars, the operation "**addition**" is performed. To obtain the total number of stars in two boxes is to perform the operation "______."

addition

8
Counting the total number of stars in the following two boxes, | ☆ ☆ ☆ | and | ☆ ☆ ☆ ☆ ☆ |, is performing the operation "______."

addition

9
In the **expression** 7 + 5, the symbol "+" indicates the operation *addition* is to be performed on the numbers 7 and 5. What operation is to be performed on the numbers 9 and 3 in the expression 9 + 3?

addition

10
6 + 7 is the expression used to indicate addition of the numbers 6 and 7. Write the expression to indicate addition of the numbers 8 and 4.

8 + 4

11
Write an expression indicating that 5 and 9 are to be added.

5 + 9

12
8 + 3 is the expression to indicate that 8 and 3 are to be added. 9 + 6 is the ______ to indicate 9 and 6 are to be added.

expression

13
3 + 8 is the ______ to indicate that the numbers 3 and 8 are to be added.

expression

14
The result of adding 9 and 7 is 16. 16 is the **sum** of the expression 9 + 7. What is the sum of the expression 5 + 8?

13

15
What is the sum of the expression 4 + 6?

10

16
The expression $7 + 4$ has a sum of 11.
$9 + 5$ has a ______ of 14. sum

17
14 is the ______ of $8 + 6$. sum

18
To **evaluate the expression** $4 + 5$ is to find the sum.
9 is the evaluation of the expression $4 + 5$. What is the evaluation of the expression $8 + 7$? 15

19
For the expression $4 + 8$, find the sum (evaluate). 12

20
Evaluate the expression. $13 + 8$ 21

21
Evaluate the expression. $4 + 9$ 13

22
Evaluate. $11 + 6$ 17

23
In evaluating the expression $(8 + 5) + 3$, the **parentheses** indicate that the first step is to find the sum of 8 and 5.
Place parentheses in $4 + 9 + 3$ to indicate that the first step is to add 4 and 9. $(4 + 9) + 3$

24
Place the parentheses in $5 + 3 + 7$ to indicate that the first step is to find the sum of 5 and 3. $(5 + 3) + 7$

25
In the expression $(8 + 5) + 4$ the parentheses indicate that the first step is to evaluate $8 + 5$. In the expression $(9 + 3) + 8$ the parentheses indicate that the first step is to evaluate ______. $9 + 3$

26
In the expression $(8 + 4) + 7$ the parentheses indicate that the first step is to evaluate ______.

$8 + 4$

27
In the expression $(3 + 9) + 6$ the parentheses indicate that the first step is to evaluate ______.

$3 + 9$

28
How many numbers are in the parentheses of the expression $(5 + 3) + 8$?

Two

29
How many numbers are added in the first step in evaluating the expression $(7 + 6) + 4$?

Two

30
In evaluating the expression $5 + 3 + 8 + 2$, place parentheses to indicate that the first step is to add 5 and 3.

$(5 + 3) + 8 + 2$

31
$12 + 7$ is the result of the first step in evaluating $(8 + 4) + 7$. What is the result of the first step in evaluating $(9 + 7) + 4$?

$16 + 4$

32
What is the result of the first step in evaluating $(8 + 5) + 9$?

$13 + 9$

33
In evaluating the expression $(4 + 3) + 5$, the result of the first step is $7 + 5$. For the second step, the two numbers 7 and 5 are added. In evaluating the expression $(5 + 4) + 2$, the result of the first step is $9 + 2$. For the second step, the two numbers 9 and ______ are added.

2

34

To evaluate the expression $(3 + 4) + 5 + 9$, two numbers are added in each step. The way to evaluate the expression is shown at the right.

$$(3 + 4) + 5 + 9$$
$$(7 + 5) + 9$$
$$12 + 9$$
$$21$$

How many numbers are added in each step? Two

35

Addition is an operation that is applied to **two** counting numbers. Whenever the sum of more than two counting numbers is to be found, how many numbers are added in each step? Two

36

Parentheses are used in the expression $(3 + 5) + 4$ to indicate that 3 is to be added to 5. The parentheses in $(7 + 6) + 8$ indicate that 7 is to be added to ______. 6

37

In the expression $4 + (7 + 6)$ the parentheses indicate the numbers to be added first. The first step in evaluating $4 + (7 + 6)$ is to add 7 and 6. The parentheses in $9 + (3 + 8)$ indicate that the first step is to add ______ and 8. 3

38

In the expression $4 + (3 + 7)$ the number 3 is to be added to the number ______ in the first step. 7

39

What number is to be added to 8 in evaluating $(8 + 9) + 4$? 9

40

What number is to be added to 9 in evaluating the expression $4 + (9 + 3)$? 3

41
Place parentheses in 9 + 3 + 6 to indicate that the first step is to add 3 and 6.

9 + (3 + 6)

42
Place parentheses in 5 + 2 + 8 to indicate that the first step is to add 5 and 2.

(5 + 2) + 8

43
Place parentheses in 7 + 1 + 13 to indicate that the first step is to add 1 and 13.

7 + (1 + 13)

44
Square brackets, [ ], can be used in the same manner as parentheses. The expression (5 + 6) + 7 can be written [5 + 6] + 7. Write the expression (2 + 7) + 6 using square brackets in place of parentheses.

[2 + 7] + 6

45
Use square brackets, [ ], to write the expression (4 + 9) + 7.

[4 + 9] + 7

46
In the expression 4 + [7 + 9] the square brackets indicate that the number 9 is to be added to the number ______.

7

47
Use square brackets to indicate that 4 is to be added to 7 in the expression 4 + 7 + 8.

[4 + 7] + 8

48
Use square brackets in the expression 4 + 5 + 7 to indicate that 5 should be added to 7 in the first step.

4 + [5 + 7]

49

Only two counting numbers can be added at one time. Whenever the sum of more than two numbers is to be found, the parentheses or square brackets are used to indicate which ______ numbers are to be added first.

two

50

In the expression $[9 + (4 + 7)] + 6$, how many numbers are in the square brackets?

three

51

In the expression $[(4 + 3) + 5] + 7$, there are three numbers in the square brackets and two numbers in the parentheses. Since only two numbers can be added at one time, the numbers in the parentheses are added first. In evaluating the expression $[(4 + 3) + 5] + 7$, 3 is added to ______ in the first step.

4

52

In the expression $2 + ([6 + 8] + 9)$, there are three numbers in the parentheses and only two numbers in the square brackets. Since only two numbers can be added at one time, the first step in evaluating the expression is to add 6 and ______.

8

Parentheses, ( ), and square brackets, [ ], are two symbols used in algebra to show the grouping of numbers in an expression. In evaluating an expression containing both parentheses and square brackets, the "innermost" grouping symbol should be evaluated first.

For example, the expression [5 + (3 + 4)] has parentheses inside the square brackets. Therefore, the expression (3 + 4) should be evaluated first, as follows:

$$[5 + (3 + 4)]$$
$$[5 + 7]$$
$$12$$

The expression ([8 + 7] + 9) has the square brackets inside the parentheses, and [8 + 7] is to be evaluated first, as follows:

$$([8 + 7] + 9)$$
$$(15 + 9)$$
$$24$$

53
In the expression 9 + [8 + (3 + 4)], 4 is added to ______ in the first step.

3

54
In the expression (4 + [7 + 9]) + 6, 9 is added to ______ in the first step.

7

55
The following steps are used to evaluate [(4 + 6) + 4] + 8:

$$[(4 + 6) + 4] + 8$$
$$[10 + 4] + 8$$
$$14 + 8$$
$$22$$

How many numbers were added in each step?

two

56

Since only two numbers can be added at a time, $(7 + [9 + 6]) + 3$ is evaluated in the following steps:

$$(7 + [9 + 6]) + 3$$
$$(7 + 15) + 3$$
$$22 + 3$$
$$25$$

Evaluate step by step. $(3 + [6 + 4]) + 9$

$(3 + 10) + 9$
$13 + 9$
22

57

Evaluate. $5 + [9 + (3 + 6)]$

$5 + [9 + 9]$
$5 + 18$
23

58

Evaluate. $([9 + 3] + 6) + 4$

$(12 + 6) + 4$
$18 + 4$
22

59

Evaluate $9 + (4 + 8)$. Evaluate $(9 + 4) + 8$. Do these expressions have the same evaluation?

Yes

60

Do the two expressions $(13 + 8) + 4$ and $13 + (8 + 4)$ have the same evaluation?

Yes

61

Do the expressions $11 + 7$ and $7 + 11$ have the same evaluation?

Yes

62

Do the expressions $[7 + 2] + 3$ and $7 + [2 + 3]$ have the same evaluation?

Yes

63
Evaluate ([5 + 8] + 3) + 7 and 5 + (8 + [3 + 7]). Do these expressions have the same evaluation?

Yes

64
Do the expressions 45 + 13 and 13 + 45 have the same evaluation?

Yes

65
Do the expressions 7 + [16 + (3 + 2)] and [7 + 16] + (3 + 2) have the same evaluation?

Yes

FEEDBACK UNIT 2

This quiz reviews the preceding unit. Answers are at the back of the book.

1. How many counting numbers can be added at one time?
2. Do the two expressions 8 + 3 and 3 + 8 have the same evaluation?
3. Do the two expressions (4 + 7) + 6 and 4 + (7 + 6) have the same evaluation?
4. Do the two expressions 9 + (4 + 1) and (9 + 4) + 1 have the same evaluation?
5. Do the two expressions (5 + 9) + [4 + 7] and [(5 + 9) + 4] + 7 have the same evaluation?
6. Do the two expressions [9 + (3 + 6)] + 2 and 9 + [(3 + 6) + 2 have the same evaluation?
7. Do the two expressions 15 + 18 and 18 + 15 have the same evaluation?
8. Do the two expressions (4 + 8) + 13 and 4 + (8 + 13) have the same evaluation?

UNIT 3: EVALUATING SUBTRACTION EXPRESSIONS

The following mathematical terms are crucial to an understanding of this unit.

Subtract Difference

1
Subtract 3 from 9. 6

2
Subtract 5 from 8. 3

3
"Subtract 4 from 7" can be written as the expression 7 – 4. Write the expression to indicate "subtract 6 from 9." 9 – 6

4
Write the expression to indicate "subtract 5 from 13." 13 – 5

5
"Subtract 4 from 11" can be written as the expression ______. 11 – 4

6
Evaluate the expression 7 – 3 by subtracting 3 from 7. 4

7
Evaluate the expression 87 – 13 by subtracting 13 from 87. 74

8
Evaluate (find the **difference**) the expression. 13 – 2 — 11

9
Evaluate the expression. 47 – 21 — 26

10
Evaluate. 251 – 127 — 124

11
What number added to 6 will give a sum of 10? — 4

12
What number added to 18 will give a sum of 31? — 13

13
What number can be filled in the blank to give the indicated sum: ______ + 4 = 12? — 8

14
Fill in the blank: ______ + 7 = 12. — 5

15
8 – 3 = ______ and ______ + 3 = 8. — 5, 5

16
24 – 15 = ______ and ______ + 15 = 24. — 9, 9

17
Every subtraction problem can be written as an addition question. 7 – 3 = ______ can be written as ______ + 3 = 7. Write 13 – 2 = ______ as an addition question. — ______ + 2 = 13

18
Write 47 – 13 = ______ as an addition question. — ______ + 13 = 47

19
Write 14 – 9 = ______ as an addition question. — ______ + 9 = 14

20

3 – 5 = ______ can be written as ______ + 5 = 3. Since there is no counting number that can be added to 5 to give a sum of 3, the blanks cannot be filled in with a counting number. Can the following blank be filled in with a counting number? 2 – 7 = ______

No

21

Using the set of counting numbers, the problem 3 – 8 = ______ has no answer. Does the problem 4 – 12 = ______ have an answer in the set of counting numbers?

No

22

Does the problem 2 – 11 = ______ have an answer in the set of counting numbers?

No

23

Is there a counting number that will answer the problem 5 – 14 = ______ correctly?

No

24

Subtraction of counting numbers is possible only when the first number is larger than the second. Is it possible to do the problem 9 – 5 = ______ with a counting number?

Yes

25

Is there a counting-number answer for 12 – 45 = ______?

No

26

Is there a counting-number answer for 47 – 93 = ______?

No

27

The problem 5 – 7 = ______ does not have a ______-number answer.

counting

28
Is there a counting-number answer for both of the following?
4 – 11 = ______ and 11 – 4 = ______

No

29
Is there a counting-number answer for both of the following?
27 – 13 = ______ and 13 – 27 = ______

No

30
Which of the following problems cannot be answered by a counting number?

11 – 4 = ______
4 – 11 = ______
11 + 4 = ______
4 + 11 = ______

4 – 11 = ______

31
Which of the following problems cannot be answered by a counting number?

47 – 13 = ______
13 + 47 = ______
47 + 13 = ______
13 – 47 = ______

13 – 47 = ______

32
Which of the following problems cannot be answered by a counting number?

78 + 53 = ______
53 + 78 = ______
53 – 78 = ______
78 – 53 = ______

53 – 78 = ______

33
Which of the following problems cannot be answered by a counting number?

89 + 64 = ______
89 – 64 = ______
64 + 89 = ______
64 – 89 = ______

64 – 89 = ______

34
Is the sum of 4 and 9 in the set of counting numbers {1, 2, 3, . . . }?

Yes

35
If any two numbers in the set of counting numbers, {1, 2, 3, . . . }, are added, is their sum also in the set?

Yes

36
Is the evaluation of 14 – 5 in the set {1, 2, 3, . . . }?

Yes

37
Is the evaluation of 3 – 15 in the set {1, 2, 3, . . . }?

No

38
Are there subtraction problems that have no answer in the set of counting numbers {1, 2, 3, . . . }?

Yes

39
As in addition, subtraction is an operation on only two numbers at a time. In the expression (13 – 3) – 4 the parentheses indicate that 3 is to be subtracted from ______.

13

40
The parentheses in (15 – 7) – 5 indicate that 7 is to be subtracted from 15. Insert parentheses in 13 – 3 – 1 to indicate that 3 is to be subtracted from 13.

(13 – 3) – 1

41
Insert parentheses in 15 – 4 – 2 to indicate that 2 is to be subtracted from 4.

15 – (4 – 2)

42
To evaluate (15 – 9) – 4, first evaluate 15 – 9 as indicated by the parentheses. What is evaluated first in the expression (18 – 3) – 5?

18 – 3

43
What is evaluated first in the expression (27 – 8) – 5?

27 – 8

44
In evaluating (14 – 6) – 2, the first step is to evaluate 14 – 6. Since 14 – 6 = 8, the result of the first step is 8 – 2. What is the result of the first step in evaluating the expression (9 – 3) – 5?

6 – 5

45
What is the result of the first step in evaluating the expression (17 – 11) – 4?

6 – 4

46
In evaluating the expression (14 – 4) – 3, the following steps are used:

(14 – 4) – 3
10 – 3
7

Evaluate the expression. (17 – 5) – 6

12 – 6 = 6

47
Evaluate the expression. (9 – 3) – 4

6 – 4 = 2

48
To evaluate 15 – (9 – 4), first evaluate 9 – 4 as indicated by the parentheses. What is evaluated first in the expression 14 – (6 – 3)?

6 – 3

49
What is evaluated first in the expression $15 - (9 - 4)$? — $9 - 4$

50
In evaluating $17 - (5 - 3)$, the first step is to evaluate $5 - 3$. Since $5 - 3 = 2$, the result of the first step is $17 - 2$. What is the result of the first step in evaluating $14 - (2 - 1)$? — $14 - 1$

51
What is the result of the first step in evaluating the expression $23 - (14 - 5)$? — $23 - 9$

52
In evaluating the expression $17 - (11 - 3)$, the following steps are used:

$$17 - (11 - 3)$$
$$17 - 8$$
$$9$$

Evaluate the expression. $19 - (12 - 7)$ — $19 - 5 = 14$

53
Evaluate the expression. $13 - (7 - 4)$ — $13 - 3 = 10$

54
Evaluate. $17 - (12 - 9)$ — $17 - 3 = 14$

55
Evaluate. $(17 - 4) - 8$ — $13 - 8 = 5$

56
Evaluate. $15 - (8 - 5)$ — 12

57
$(10 - 3) - 2$ has an evaluation of 5, but $10 - (3 - 2) = 9$. Do the expressions $(14 - 8) - 5$ and $14 - (8 - 5)$ have the same evaluation? — No

58
Do (15 – 3) – 2 and 15 – (3 – 2) have the same evaluation?

No

59
Do the expressions (18 – 9) – 6 and 18 – (9 – 6) have the same evaluation?

No

60
By placing the parentheses in different positions in the expression 14 – 6 – 3, different evaluations are obtained. (14 – 6) – 3 = 5 and 14 – (6 – 3) = 11. Do (19 – 8) – 5 and 19 – (8 – 5) have the same evaluation?

No

61
Do (14 – 7) – 5 and 14 – (7 – 5) have the same evaluation?

No

62
Do (15 – 7) – 4 and 15 – (7 – 4) have the same evaluation?

No

FEEDBACK UNIT 3

This quiz reviews the preceding unit. Answers are at the back of the book.

1. Do 3 + 9 and 9 + 3 have the same evaluation?
2. Do 9 – 7 and 7 – 9 have the same evaluation?
3. Do (15 + 9) + 3 and 15 + (9 + 3) have the same evaluation?
4. Do (15 – 9) – 3 and 15 – (9 – 3) have the same evaluation?
5. Evaluate. 15 – (9 – 2)
6. Evaluate. 24 – (17 – 5)
7. Evaluate. [26 – 9] – 11
8. Evaluate. 19 – [8 – 3]

UNIT 4: EVALUATING MULTIPLICATION EXPRESSIONS

The following mathematical terms are crucial to an understanding of this unit.

Total
Product
Multiplication
Multiplication Expression

1
How many boxes with 4 stars each are there?
☆ ☆ ☆ ☆ | ☆ ☆ ☆ ☆ | ☆ ☆ ☆ ☆

3

2
How many stars are there in each of the 3 boxes?
☆ ☆ ☆ ☆ | ☆ ☆ ☆ ☆ | ☆ ☆ ☆ ☆

4

3
In frame 2 each box had the same number of stars.
What is the **total** number of stars in all 3 boxes?

12

4
How many boxes are there below? How many stars are there in each box?
☆ ☆ | ☆ ☆ | ☆ ☆ | ☆ ☆ | ☆ ☆

5 boxes, 2 stars in each box

5
What is the total number of stars in the 5 boxes in frame 4?

10

6
What is the total number of stars in the following boxes?
☆ ☆ ☆ ☆ ☆ | ☆ ☆ ☆ ☆ ☆ | ☆ ☆ ☆ ☆ ☆

15

7
What is the total number of stars
in the following 3 boxes?

☆ ☆ ☆ ☆ ☆ ☆ ☆ ☆	☆ ☆ ☆ ☆ ☆ ☆ ☆ ☆
☆ ☆ ☆ ☆ ☆ ☆ ☆ ☆	

24

8
What is the total number of stars in 4
boxes if each box has 5 stars?

20

9
Find the total number of stars in 5 boxes
if each box has 9 stars.

45

10
Find the total number of stars in 12 boxes
if each box has 4 stars.

48

11
To find the total number of stars in 4 boxes if each
box has 6 stars, the expression $6 + 6 + 6 + 6$ can be
used. Write the expression to show the total number
of stars in 3 boxes if each box has 7 stars.

$7 + 7 + 7$

12
Write the expression to show the total number
of stars in 4 boxes if each box has 5 stars.

$5 + 5 + 5 + 5$

13
Write the expression to show the total number
of stars in 3 boxes if each box has 15 stars.

$15 + 15 + 15$

14
$6 + 6 + 6 + 6$ is the expression to show the total
number of stars in 4 boxes if each box has 6 stars.
$6 + 6 + 6 + 6$ can be shown by the expression 4×6
using the operation **multiplication**. What is the
multiplication expression for $5 + 5 + 5 + 5$?

4×5

15
What is the multiplication expression for $9 + 9 + 9 + 9 + 9$? | 5×9

16
What is the multiplication expression for $27 + 27 + 27 + 27$? | 4×27

17
Write $4 + 4 + 4 + 4 + 4 + 4$ as a multiplication expression. | 6×4

18
Since $3 + 3 + 3 + 3 + 3 = 15$, then $5 \times 3 = 15$. Since $4 + 4 + 4 = 12$, then $3 \times 4 =$ ______. | 12

19
Since $4 + 4 + 4 + 4 + 4 = 20$, then $5 \times 4 = 20$. Since $8 + 8 + 8 = 24$, then $3 \times 8 =$ ______. | 24

20
Since $7 + 7 + 7 + 7 + 7 + 7 + 7 + 7 = 56$, then $8 \times 7 =$ ______. | 56

21
The evaluation of a multiplication expression is its **product**. The product of 4×5 is 20. 32 is the ______ of 4×8. | product

22
The product of 6×5 is 30. The product of 9×5 is ______. | 45

23
Evaluate the expression. 3×13 | 39

24
$8 \times 15 =$ ______ | 120

In algebra the multiplication symbol × is not used, because the letter x is frequently used as a variable. To avoid the possibility of confusion between meaning multiplication or representing a variable, a raised dot (•) will be used to represent multiplication throughout the remainder of the text.

25
In mathematics a raised dot (•) replaces the symbol (×) to indicate the operation multiplication. 3×4 is written $3 \cdot 4$. Using the dot to replace the (×), write 5×8.

$5 \cdot 8$

26
8×6 is written $8 \cdot 6$. Using the dot symbol, write 7×4.

$7 \cdot 4$

27
$6 \cdot 3 =$ ______

18

28
The product of the expression $10 \cdot 3$ is ______.

30

29
The product of the expression $2 \cdot 14$ is ______.

28

30
As in addition and subtraction, multiplication can be performed with only two numbers at a time. How many counting numbers may be multiplied at one time?

two

31
In evaluating the expression $(8 \cdot 2) \cdot 3$, the first step is to find the product of $8 \cdot 2$. The first step in evaluating the expression $(4 \cdot 5) \cdot 7$ is to find the product of ______.

$4 \cdot 5$

32

In evaluating the expression $5 \cdot (4 \cdot 8)$, the first step is to find the product of ______.

$4 \cdot 8$

33

In evaluating the expression $(3 \cdot 2) \cdot 5$, 3 and 2 are multiplied first. Which numbers are to be multiplied first in evaluating the expression $(5 \cdot 3) \cdot 6$?

5 and 3

34

In the expression $(4 \cdot 7) \cdot 3$ there are three counting numbers. Which two are to be multiplied first?

4 and 7

35

The following steps are used in evaluating $4 \cdot (3 \cdot 2)$:

$$4 \cdot (3 \cdot 2)$$
$$4 \cdot 6$$
$$24$$

Evaluate step by step the expression. $5 \cdot (3 \cdot 4)$

$5 \cdot 12$
60

36

Evaluate. $2 \cdot [5 \cdot 3]$

$2 \cdot 15 = 30$

37

Evaluate. $(3 \cdot 2) \cdot 6$

$6 \cdot 6 = 36$

38

To evaluate the expression $(3 \cdot 7) \cdot 4$, the following steps are used:

$$(3 \cdot 7) \cdot 4$$
$$21 \cdot 4$$
$$84$$

Evaluate step by step the expression. $(3 \cdot 5) \cdot 4$

$15 \cdot 4 = 60$

39

Evaluate the expression. $[5 \cdot 7] \cdot 3$

$35 \cdot 3 = 105$

40
Evaluate. $4 \cdot (5 \cdot 6)$

$4 \cdot 30 = 120$

41
In the expression $(10 \cdot 4) \cdot [7 \cdot 2]$ the 10 is multiplied by the 4 and the 7 is multiplied by the 2. In the expression $[8 \cdot 1] \cdot (6 \cdot 2)$ the 8 is multiplied by the 1 and the 6 is multiplied by the ______.

2

42
In the expression $[9 \cdot 6] \cdot (3 \cdot 7)$ the parentheses indicate that the 3 is to be multiplied by the 7. Which number is to be multiplied by the 5 in the expression $[4 \cdot 7] \cdot (5 \cdot 8)$?

8

43
To evaluate the expression $(5 \cdot 4) \cdot (3 \cdot 2)$, the following steps are used:

$$(5 \cdot 4) \cdot (3 \cdot 2)$$
$$20 \cdot 6$$
$$120$$

Evaluate step by step the expression. $(2 \cdot 5) \cdot (4 \cdot 6)$

$10 \cdot 24 = 240$

44
Evaluate the expression. $[3 \cdot 5] \cdot (4 \cdot 2)$

$15 \cdot 8 = 120$

45
Evaluate. $[10 \cdot 4] \cdot [3 \cdot 5]$

$40 \cdot 15 = 600$

46
The expression $5 \cdot ([2 \cdot 4] \cdot 7)$ has three counting numbers in the parentheses and two numbers in the square brackets. $[2 \cdot 4]$ is the innermost grouping and $2 \cdot 4$ is performed first. Which numbers are to be multiplied first in evaluating the expression $8 \cdot ([3 \cdot 9] \cdot 5)$?

3 and 9

47

The first step in evaluating $6 \cdot [5 \cdot (4 \cdot 8)]$ is to find the product of 4 and 8. Which two numbers are to be multiplied first in evaluating the expression $2 \cdot [3 \cdot (4 \cdot 5)]$?

4 and 5

48

Which two numbers are to be multiplied first in evaluating the expression $5 \cdot ([2 \cdot 4] \cdot 3)$?

2 and 4

49

In the expression $[2 \cdot (3 \cdot 7)] \cdot 5$, which two numbers are to be multiplied first?

3 and 7

50

The following steps are used to evaluate $[(4 \cdot 3) \cdot 5] \cdot 6$:

$$[(4 \cdot 3) \cdot 5] \cdot 6$$

$$[12 \cdot 5] \cdot 6$$

$$60 \cdot 6$$

$$360$$

Evaluate step by step. $[(2 \cdot 4) \cdot 3] \cdot 6$

$[8 \cdot 3] \cdot 6$
$24 \cdot 6$
144

51

Evaluate the expression. $(5 \cdot [3 \cdot 2]) \cdot 4$

120

52

Evaluate. $9 \cdot [1 \cdot (3 \cdot 2)]$

54

53

Evaluate. $[7 \cdot 5] \cdot [3 \cdot 2]$

210

54

Do the following expressions have the same evaluation?
$(5 \cdot 3) \cdot 4$ and $5 \cdot (3 \cdot 4)$?

Yes, 60

55
Evaluate the expressions [(4 • 3) • 6] • 5 and 4 • [3 • (6 • 5)]. Are the evaluations the same?

Yes, 360

56
Evaluate the expressions [(4 • 8) • 6] • 5 and 4 • [8 • (6 • 5)]. Are the evaluations the same?

Yes

57
Evaluate the expressions 6 • 5 and 5 • 6. Are the evaluations the same?

Yes

58
Evaluate the expressions (2 • 3) • (6 • 5) and [6 • 5] • (2 • 3). Are the evaluations the same?

Yes

59
Evaluate the expressions [(6 • 2) • 3] • 4 and 6 • [2 • (3 • 4)]. Are the evaluations the same?

Yes

FEEDBACK UNIT 4

This quiz reviews the preceding unit. Answers are at the back of the book.

1. How many counting numbers can be multiplied at one time?
2. Evaluate. (8 • 2) • 3
3. Evaluate. 5 • [4 • 7]
4. Do (7 • 3) • 2 and 7 • (3 • 2) have the same evaluation?
5. Do 9 • 6 and 6 • 9 have the same evaluation?
6. Evaluate. 3 • (8 • 4)
7. Evaluate. [2 • (5 • 3)] • 4
8. Evaluate. 3 • [4 • (4 • 1)]

UNIT 5: EVALUATING DIVISION EXPRESSIONS

The following mathematical terms are crucial to an understanding of this unit.

Division Quotient

1

$16 \div 4$ is read "16 divided by 4."
Evaluate $16 \div 4$.

4

2

$15 \div 3 =$ ______

5

3

$27 \div 9 =$ ______

3

4

$21 \div 7 =$ ______

3

5

Every **division** problem is a multiplication question. The problem $15 \div 3 = ?$ asks "What number can be multiplied by 3 to give a product of 15?" The problem $27 \div 9 = ?$ asks "What number can be multiplied by 9 to give a product of ______?"

27

6

The problem $72 \div 9 = ?$ asks "What number can be multiplied by 9 to give 72?" The problem $36 \div 4 = ?$ asks "What number can be multiplied by 4 to give ______?"

36

7

The division problem $20 \div 5 =$ ______ may be written as the multiplication question $5 \bullet$ ______ $= 20$. Write $18 \div 9 =$ ______ as a multiplication question.

$9 \bullet$ ______ $= 18$

8

Write $63 \div 7 =$ ______ as a multiplication question.

$7 \bullet$ ______ $= 63$

9

Write $48 \div 6 =$ ______ as a multiplication question.

$6 \bullet$ ______ $= 48$

The division problem $27 \div 6 =$ ______ does not have a counting-number answer because there is no counting number that can be multiplied by 6 to give a product of 27.

The division problem $30 \div 6 =$ ______ has the answer 5, because $5 \bullet 6 = 30$.

Some division problems have counting-number answers and others do not, as shown by the following examples.

10

$13 \div 3 =$ ______ can be written ______ $\bullet 3 = 13$. There is no counting number answer for either problem. Does the problem $12 \div 7 =$ ______ have a counting-number answer?

No

11

Using the set of counting numbers, the division problem $18 \div 7 =$ ______ has no answer. Does $35 \div 4 =$ ______ have a counting-number answer?

No

12

$41 \div 5 =$ ______ is a problem that cannot be answered in the set of counting numbers $\{1, 2, 3, \ldots\}$. There are many division problems that do not have a counting number for an answer. Does $17 \div 4 =$ ______ have a counting-number answer?

No

13

Is there a counting-number answer to both of the following problems?

$18 \div 6 =$ ______ and $6 \div 18 =$ ______

No

14

Which of the following problems cannot be answered with a counting number?

$9 \bullet 3 =$ ______
$9 \div 3 =$ ______
$3 \bullet 9 =$ ______
$3 \div 9 =$ ______

$3 \div 9 =$ ______

15

Which of the following problems does not have an answer in the set of counting numbers {1, 2, 3, . . . }?

$24 \bullet 6 =$ ______
$6 \div 24 =$ ______
$24 \div 6 =$ ______
$6 \bullet 24 =$ ______

$6 \div 24 =$ ______

16

Which of the following problems does not have an answer in the set of counting numbers?

$32 \div 8 =$ ______
$8 \bullet 32 =$ ______
$8 \div 32 =$ ______
$32 \bullet 8 =$ ______

$8 \div 32 =$ ______

17

Which of the following problems cannot be answered using the set of counting numbers?

$150 \bullet 5 =$ ______
$5 \bullet 150 =$ ______
$150 \div 5 =$ ______
$5 \div 150 =$ ______

$5 \div 150 =$ ______

18
The product of 4 and 7 is 28. Is the product of two counting numbers always in the set of counting numbers?

Yes

19
The **quotient** (answer) of $28 \div 4$ is 7. Is the quotient of $17 \div 3$ a counting number?

No

20
Do all division problems of counting numbers have an answer in the set of counting numbers?

No

21
As in addition, subtraction, and multiplication, division can only be performed with two numbers at a time. In the expression $(24 \div 8) \div 3$, the parentheses indicate that 24 is to be divided by ______.

8

22
In the expression $24 \div (8 \div 4)$, 8 is to be divided by ______.

4

23
In the expression $24 \div 8 \div 2$, insert parentheses to indicate that 8 is to be divided by 2.

$24 \div (8 \div 2)$

24
In the expression $80 \div 4 \div 2$, insert square brackets to indicate that 80 is to be divided by 4.

$[80 \div 4] \div 2$

25
To evaluate $(72 \div 4) \div 9$, first evaluate $72 \div 4$ as indicated by the parentheses. What is evaluated first in the expression $(40 \div 2) \div 5$?

$40 \div 2$

26

In evaluating $(18 \div 2) \div 3$, the first step is to evaluate $18 \div 2$. Since $18 \div 2 = 9$, the result of the first step is $9 \div 3$. What is the result of the first step in evaluating $(40 \div 5) \div 4$?

$8 \div 4$

27

What is the result of the first step in evaluating $[60 \div 3] \div 4$?

$20 \div 4$

28

In evaluating the expression $(24 \div 2) \div 6$, the following steps are used:

$$(24 \div 2) \div 6$$
$$12 \div 6$$
$$2$$

Evaluate the expression. $(60 \div 3) \div 5$

$20 \div 5 = 4$

29

Evaluate the expression. $(48 \div 4) \div 4$

$12 \div 4 = 3$

30

In evaluating the expression $24 \div (8 \div 2)$, the following steps are used:

$$24 \div (8 \div 2)$$
$$24 \div 4$$
$$6$$

Evaluate. $40 \div (6 \div 3)$

$40 \div 2 = 20$

31

Evaluate. $18 \div (36 \div 6)$

$18 \div 6 = 3$

32

Evaluate. $8 \div [4 \div 2]$

$8 \div 2 = 4$

33

Evaluate. $[90 \div 3] \div 5$

$30 \div 5 = 6$

34
Does $(8 \div 4) \div 2$ have the same evaluation as $8 \div (4 \div 2)$?

No

35
Do $(40 \div 4) \div 2$ and $40 \div (4 \div 2)$ have the same evaluation?

No

36
By placing parentheses differently in the expression $50 \div 10 \div 5$, two different evaluations are obtained: $(50 \div 10) \div 5 = 1$ and $50 \div (10 \div 5) = 25$. Do $(45 \div 15) \div 3$ and $45 \div (15 \div 3)$ have the same evaluation?

No

37
Do the expressions $60 \div (6 \div 2)$ and $(60 \div 6) \div 2$ have the same evaluation?

No

FEEDBACK UNIT 5

This quiz reviews the preceding unit. Answers are at the back of the book.

1. Do $3 \cdot 8$ and $8 \cdot 3$ have the same evaluation?
2. Do $8 \div 4$ and $4 \div 8$ have the same evaluation?
3. Do $(8 \cdot 5) \cdot 2$ and $8 \cdot (5 \cdot 2)$ have the same evaluation?
4. Do $(8 \div 4) \div 2$ and $8 \div (4 \div 2)$ have the same evaluation?
5. Evaluate. $(12 \div 3) \div 4$
6. Evaluate. $24 \div (72 \div 9)$
7. Evaluate. $18 \div [12 \div 4]$
8. Evaluate. $(84 \div 3) \div 7$

UNIT 6: EVALUATING NUMERICAL EXPRESSIONS INVOLVING MORE THAN ONE OPERATION

The following mathematical terms are crucial to an understanding of this unit.

Numerical expression
Order of Operations
Evaluating numerical expressions

Because subtraction and division problems may not have counting-number answers, this unit deals only with expressions involving addition and multiplication.

1
An expression involving addition and/or multiplication is a **numerical expression**.
5 + (3 • 4) is a numerical expression.
7 + (5 • 2) is a ______ expression. — numerical

2
(5 • 3) + 6 is a ______ expression. — numerical

3
3 • [4 + 5] is a ______ expression. — numerical

4
4 + 7 and 3 • 9 are numerical expressions.
5 + 9 is a ______ expression. — numerical

5
Write a numerical expression to indicate the product of 12 and 7. — 12 • 7

6
Write a numerical expression to indicate the sum of 11 and 16. — 11 + 16

7
Numerical expressions may involve one or more operations. Is $4 \bullet (3 + 6)$ a numerical expression?

Yes

8
To **evaluate the numerical expression** $(2 + 3) \bullet 4$, the parentheses indicate that the first step is to evaluate $2 + 3$. The first step in evaluating $(7 + 4) \bullet 3$ is to evaluate ______.

$7 + 4$

9
The first step in evaluating $(5 + 4) \bullet 7$ is to evaluate ______.

$5 + 4$

10
To evaluate $3 + (6 \bullet 5)$, the first step is to evaluate $6 \bullet 5$ as indicated by the parentheses. The first step in evaluating $4 + (7 \bullet 9)$ is to evaluate ______.

$7 \bullet 9$

11
The first step in evaluating $12 + (8 \bullet 5)$ is to evaluate ______.

$8 \bullet 5$

12
To evaluate $5 + (3 \bullet 7)$, the following steps are used:

$$5 + (3 \bullet 7)$$
$$5 + 21$$
$$26$$

Evaluate. $7 + (2 \bullet 6)$

$7 + 12 = 19$

13
Evaluate the expression. $6 + [8 \bullet 3]$

$6 + 24 = 30$

14
Evaluate. $(7 \bullet 3) + 3$

$21 + 3 = 24$

15

To evaluate $(4 + 3) \cdot 2$, the following steps are used:

$$(4 + 3) \cdot 2$$
$$7 \cdot 2$$
$$14$$

Evaluate. $(3 + 6) \cdot 5$

$9 \cdot 5 = 45$

16

Evaluate the expression. $[9 + 3] \cdot 4$

$12 \cdot 4 = 48$

17

Evaluate. $3 \cdot (7 + 1)$

$3 \cdot 8 = 24$

18

In $5 + [3 \cdot (4 + 2)]$, how many numbers are in the square brackets?

three

19

In $5 + [3 \cdot (4 + 2)]$, how many numbers are in the parentheses?

two

20

In addition only two numbers can be added at one time. In multiplication only two numbers can be multiplied at one time. In either addition or multiplication only ______ numbers can be combined at one time.

two

21

There are only two numbers in the parentheses of the expression $5 + [3 \cdot (4 + 2)]$ and the first step is to evaluate $4 + 2$. What is the first step in evaluating $8 + [9 \cdot (3 + 1)]$?

$3 + 1$

22

The result of the first step in evaluating $7 + [(4 + 1) \cdot 6]$ is $7 + [5 \cdot 6]$. What is the result of the first step in evaluating $8 + [(3 + 4) \cdot 2]$?

$8 + [7 \cdot 2]$

23

The result of the first step in evaluating $[(3 \cdot 5) + 6] + 4$ is ______.

$[15 + 6] + 4$

24

The following steps are used to evaluate $[(3 \cdot 7) + 2] + 6$:

$$[(3 \cdot 7) + 2] + 6$$
$$[21 + 2] + 6$$
$$23 + 6$$
$$29$$

Evaluate. $[(6 \cdot 5) + 4] + 3$

$[30 + 4] + 3$
$34 + 3$
37

To evaluate an expression that has two or more grouping symbols, the expression in the innermost grouping symbol should be evaluated first.

25

Evaluate the expression. $[(9 \cdot 4) + 2] + 1$

$[36 + 2] + 1$
$38 + 1$
39

26

Evaluate the expression. $[(5 + 3) \cdot 4] + 2$

$[8 \cdot 4] + 2$
$32 + 2$
34

27

Evaluate. $[3 \cdot (4 + 1)] + 7$

$[3 \cdot 5] + 7$
$15 + 7$
22

28

The following steps are used to evaluate $5 + (2 + [6 \cdot 2])$:

$$5 + (2 + [6 \cdot 2])$$
$$5 + (2 + 12)$$
$$5 + 14$$
$$19$$

Evaluate. $10 + (6 + [3 \cdot 5])$ 31

29

Evaluate. $6 + (8 + [3 \cdot 4])$ 26

30

Evaluate. $8 + [(4 \cdot 1) + 5]$ 17

31

Evaluate. $9 + [(3 + 8) \cdot 2]$ 31

32

When square brackets or parentheses do not appear in an expression, **multiplication is always performed before any addition**. In the expression $3 + 4 \cdot 5$, the first step is to multiply 4 and 5. The first step in evaluating $7 + 2 \cdot 6$ is to __________. multiply 2 and 6

33

In the expression with no parentheses, $5 \cdot 7 + 2$, the multiplication is performed before the addition.

Evaluate. $5 \cdot 7 + 2$ 37

When no parentheses or square brackets appear in a numerical expression, all multiplication is to be performed before any addition. The **Order of Operations** is: Grouping symbols always indicate where the evaluation of a numerical expression is to begin. When two or more operations are included in a grouping symbol, all multiplication is to be performed before any addition.

34
In the expression $4 + 3 \cdot 5$ there are no parentheses or square brackets. The operation multiplication is to be performed before any addition. Evaluate $4 + 3 \cdot 5$.

19

35
The first step in evaluating the expression $7 + 6 \cdot 3$ is to evaluate ______.

$6 \cdot 3$

36
The first step in evaluating $5 \cdot 4 + 7$ is to evaluate ______.

$5 \cdot 4$

37
In evaluating $(8 + 2) \cdot 3$, the parentheses indicate the first step is to evaluate $8 + 2$. In the expression with no parentheses, $8 + 2 \cdot 3$, the first step is to ____________.

multiply 2 and 3

38
In $4 \cdot [3 + 6]$ the square brackets indicate the first step is to evaluate $3 + 6$. In the expression $4 \cdot 3 + 6$ the first step is to ____________.

multiply 4 and 3

39
$9 \cdot (3 + 2) = 45$, because the parentheses indicate the first step is to add 3 and 2. Evaluate $8 \cdot (7 + 3)$.

80

40
$(4 + 7) \cdot 3 = 33$, because the parentheses indicate the first step is to add 4 and 7. Evaluate $(5 + 1) \cdot 4$.

24

41
$3 \cdot (6 + 4) =$ ______

30

42
$(4 + 3) \cdot 5 =$ ______

35

43

$5 \cdot 4 + 2 = 22$, because evaluating $5 \cdot 4$ is the first step when no parentheses are in the expression. Evaluate $6 \cdot 3 + 7$.

25

44

$3 + 6 \cdot 7 = 45$, because $6 \cdot 7$ is the first step when there are no parentheses in the expression. Evaluate $5 + 6 \cdot 3$.

23

45

$9 \cdot 4 + 6 =$ ______

42

46

$5 + 9 \cdot 3 =$ ______

32

47

$8 + 5 \cdot 2 =$ ______

18

48

$3 + 5 \cdot 7 =$ ______

38

49

$(4 + 2) \cdot 5 =$ ______

30

50

$9 \cdot 3 + 12 =$ ______

39

51

$4 \cdot [8 + 1] =$ ______

36

52

$7 + 2 \cdot 9 =$ ______

25

53

Do $(3 + 4) \cdot 5$ and $3 + 4 \cdot 5$ have the same evaluation?

No

54

Do $9 + 4 \cdot 3$ and $(9 + 4) \cdot 3$ have the same evaluation?

No

55

$5 + (3 \cdot 4) = 5 + 3 \cdot 4$ is true because the parentheses in the left expression indicate to multiply 3 and 4 first, and the lack of parentheses in the right expression also indicates to multiply 3 and 4 first. Do $8 + 5 \cdot 7$ and $8 + (5 \cdot 7)$ have the same evaluation?

Yes

56

Do $3 \cdot 5 + 7$ and $[3 \cdot 5] + 7$ have the same evaluation?

Yes

57

Do $9 \cdot 6 + 4$ and $9 \cdot (6 + 4)$ have the same evaluation?

No

58

Do $5 + (3 \cdot 7)$ and $5 + 3 \cdot 7$ have the same evaluation?

Yes

59

Evaluate. $3 + 9 \cdot 2$

21

60

Evaluate. $6 \cdot 2 + 7$

19

61

Evaluate. $8 + 4 \cdot 3$

20

62

Evaluate. $5 \cdot (1 + 4)$

25

63

$(5 + 3) \cdot (2 + 4) = 8 \cdot 6 = 48$ because the parentheses indicate that 5 is to be added to 3 and 2 is to be added to 4. Evaluate $(4 + 7) \cdot (8 + 1)$.

99

64

Evaluate. $(2 + 6) \cdot [5 + 3]$

64

65

Evaluate. [7 + 2] • [8 + 2]

90

66

Because there are no parentheses in the expression 5 • 2 + 8 • 3, 5 is multiplied by 2 and 8 is multiplied by 3 to give 10 + 24 = 34.

Evaluate. 9 • 6 + 4 • 7.

54 + 28 = 82

67

Evaluate. 6 • 3 + 2 • 7

18 + 14 = 32

68

Evaluate. 10 • 3 + 8 • 5

70

69

Evaluate. 4 • 9 + 2 • 5

46

70

Evaluate. 5 • 3 + 2 • 4

23

71

The following steps are used to evaluate the expression 3 + [6 • (4 + 7)]:

3 + [6 • (4 + 7)]

3 + [6 • 11]

3 + 66

69

Evaluate. 5 + [3 • (5 + 2)]

26

72

Evaluate. 6 + [(3 + 5) • 4]

38

73

Evaluate. [6 • (3 • 3)] + 5

59

74

The following steps are used to evaluate the expression $(5 \cdot 4 + 3) + 2$:

$$(5 \cdot 4 + 3) + 2$$
$$(20 + 3) + 2$$
$$23 + 2$$
$$25$$

Evaluate $(8 \cdot 7 + 4) + 3 =$ ______. 63

75

$(4 + 9 \cdot 2) + 5 =$ ______ 27

76

$7 + [8 \cdot 2 + 1] =$ ______ 24

77

$(4 + 3 \cdot 6) + 1 =$ ______ 23

78

$2 + [4 \cdot (1 + 5)] =$ ______ 26

79

$[(9 + 2) \cdot 4] + 3 =$ ______ 47

80

$5 + [8 \cdot 3 + 2] =$ ______ 31

FEEDBACK UNIT 6

This quiz reviews the preceding unit. Answers are at the back of the book.

1. $5 + 1 \cdot 3 =$ ______
2. $3 \cdot (7 + 5) =$ ______
3. $3 + 3 \cdot 2 =$ ______
4. $2 \cdot 7 + 5 =$ ______
5. $(5 \cdot 3 + 2) \cdot 6 =$ ______
6. $8 + [(5 + 4) \cdot 2] =$ ______
7. $4 \cdot 2 + 7 =$ ______
8. $12 + 3 \cdot 9 =$ ______
9. $8 + (7 + 2 \cdot 3) =$ ______
10. $(6 + 5) \cdot 8 =$ ______
11. $(2 + 3) \cdot (6 + 4) =$ ______
12. $8 + [3 \cdot (2 \cdot 5)] =$ ______

UNIT 7: FACTORS AND MULTIPLES OF COUNTING NUMBERS

The following mathematical terms are crucial to an understanding of this unit.

Factor
Multiple
Highest common factor (HCF)
Least common multiple (LCM)

1
Factors are used with the operation multiplication. "9 is a factor of 27" means that 9 multiplied by some counting number is 27. Is 9 a factor of 36?

Yes, $9 \cdot 4 = 36$

2
If 5 is one factor of 30, what is the other factor?

6

3
If 8 is one factor of 32, what is the other factor?

4

4
Factors of 10 are 5 and 2. Factors of 21 are 7 and ______.

3

5
Factors of 28 are 4 and 7. Factors of 40 are 8 and ______.

5

6
Using counting numbers, 15 can be written as two different multiplication expressions:

$1 \cdot 15$ and $3 \cdot 5$

Using counting numbers, write two multiplication expressions for 14.

$1 \cdot 14$ or $14 \cdot 1$
$2 \cdot 7$ or $7 \cdot 2$

7
Using counting numbers, write 35 as two different multiplication expressions.

1 • 35, 5 • 7

8
Using counting numbers, write three different multiplication expressions for 18.

1 • 18, 2 • 9, 3 • 6

9
Using counting numbers, write all the multiplication expressions that have a product of 24.

1 • 24, 2 • 12,
3 • 8, 4 • 6

10
1 • 15 = 15 and 3 • 5 = 15. 1, 3, 5, and 15 are the counting-number **factors** of 15. Therefore, the set of all counting-number factors of 15 is {1, 3, 5, 15}. Write the set of all counting-number factors of 10.

{1, 2, 5, 10}

11
Write the set of all counting-number factors of 8.

{1, 2, 4, 8}

12
Write the set of all counting-number factors of 9.

{1, 3, 9}

13
Write the set of all counting-number factors of 11.

{1, 11}

14
Write the set of all counting-number factors of 12.

{1, 2, 3, 4, 6, 12}

15
Write the set of all counting-number factors of 4.

{1, 2, 4}

Every counting number has at least two factors, the counting number itself and the number 1. For example, the set of all factors of 19 is {1, 19}.

Some counting numbers have more than two factors. For example, the set of all factors of 12 is {1, 2, 3, 4, 6, 12}.

For convenience, list factors from smallest to largest when writing a set of factors.

16
The set of all factors of 12 is {1, 2, 3, 4, 6, 12}.
The set of all factors of 18 is {1, 2, 3, 6, 9, 18}.
1, 2, 3, and 6 are elements of both sets. The largest number that is a factor of both 12 and 18 is ______.

6

17
The set of all factors of 10 is {1, 2, 5, 10}.
The set of all factors of 15 is {1, 3, 5, 15}.
The numbers 1 and 5 are in both sets. The largest number that is a factor of both 10 and 15 is ______.

5

18
The set of all factors of 4 is {1, 2, 4}.
The set of all factors of 8 is {1, 2, 4, 8}.
The largest number that is a factor of both 4 and 8 is ______.

4

The **highest common factor (HCF)** of two counting numbers is the largest factor that is in both sets of factors. For example, 5 is the highest common factor (HCF) of 10 and 15, because 5 is the largest number that is a factor of both 10 and 15.

19
4 is the **highest common factor (HCF)** of 4 and 8.
Factors of 10 are: 1, 2, 5, 10
Factors of 35 are: 1, 5, 7, 35
What is the highest common factor of 10 and 35? 5

20
Factors of 6 are: 1, 2, 3, 6. Factors of 10 are: 1, 2, 5, 10.
The highest common factor (HCF) of 6 and 10 is ______. 2

21
What is the highest common factor (HCF) of 12 and 15? 3

22
What is the highest common factor (HCF) of 8 and 20? 4

23
What is the highest common factor (HCF) of 10 and 30? 10

24
What is the HCF of 6 and 7? 1

The highest common factor (HCF) of two counting numbers may be obtained by first writing all the factors of the smaller of the two numbers. Then select the largest of those factors that is also a divisor of the larger number. This number will be the highest common factor for the original two counting numbers.

25
To find the highest common factor of 12 and 20, first write all the factors of 12, because 12 is smaller than 20. The factors of 12 are: __________ 1, 2, 3, 4, 6, 12

26
The factors of 12 are: 1, 2, 3, 4, 6, 12.
To find the HCF of 12 and 20, check each factor of 12 to determine if it is also a factor of 20.
Begin with 12. Find the HCF of 12 and 20. 4

27
To find the highest common factor of 16 and 40, first write all the factors of 16. The factors of 16 are: ______________

1, 2, 4, 8, 16

28
The factors of 16 are: 1, 2, 4, 8, 16.
To find the HCF of 16 and 40, check each factor of 16 to determine if it is also a factor of 40.
Begin with 16. Find the HCF of 16 and 40.

8

29
Find the highest common factor (HCF) of 24 and 60.

12

30
Find the HCF of 18 and 54.

18

31
Find the HCF of 3 and 17.

1

32
Find the HCF of 26 and 65.

13

33
Find the HCF of 32 and 80.

16

34
The numbers 5, 10, 15, 20, 25, 30, . . . are **multiples** of 5. They are found by multiplying 5 by 1, by 2, by 3, . . . The numbers 3, 6, 9, 12, 15, 18, . . . are multiples of ______.

3

35
The numbers 9, 18, 27, 36, . . . are multiples of 9.

$1 \bullet 9 = 9$
$2 \bullet 9 = 18$
$3 \bullet 9 = 27$
$4 \bullet 9 = 36$

Find the next multiple of 9.

45 because $5 \bullet 9 = 45$

36
The first five multiples of 8 are: 8, 16, 24, 32, and 40.
List the first five multiples of 6. 6, 12, 18, 24, 30

The **least common multiple (LCM)** of two counting numbers is the smallest counting number that is a multiple of both numbers. The set of multiples of 6 is {6, 12, 18, . . .}, and the set of multiples of 9 is {9, 18, 27, . . .}. The smallest number that appears in both sets is 18. Therefore, 18 is the least common multiple (LCM) of 6 and 9.

37
The multiples of 8 are 8, 16, 24, . . .
The multiples of 12 are 12, 24, 36, . . .
The **least common multiple (LCM)** of 8 and 12 is _____, because it is the smallest number that appears in both lists. 24

38
The multiples of 6 are 6, 12, 18, . . .
The multiples of 4 are 4, 8, 12, . . .
What is the LCM of 6 and 4? 12

39
The multiples of 10 are 10, 20, 30, . . .
The multiples of 15 are 15, 30, 45, . . .
What is the least common multiple of 10 and 15? 30

40
What is the LCM of 7 and 3? 21

41
24 is the smallest counting number that is a multiple of both 3 and 8. What is the least common multiple (LCM) of 3 and 8? 24

42
What is the least common multiple (LCM) of 3 and 5? 15

43
What is the LCM of 6 and 9?

18

44
What is the LCM of 8 and 10?

40

In the next few frames we are going to use the highest common factor of two counting numbers to find their least common multiple. The least common multiple can be found by dividing the product of two numbers by their highest common factor.

45
To find the least common multiple of 12 and 18, first find their highest common factor. Then find the product of 12 and 18 and divide by the HCF.

a. the HCF of 12 and 18 is 6

b. $\frac{\overset{2}{\cancel{12}} \cdot 18}{\underset{1}{\cancel{6}}} = 2 \cdot 18 = 36$

Therefore, 36 is the least common multiple of 12 and 18.
What is the LCM of 10 and 15?

$\frac{10 \cdot 15}{5} = 30$

46
Using their HCF, find the LCM of 14 and 21.

$\frac{14 \cdot 21}{7} = 42$

47
Find the LCM of 24 and 40.

$\frac{24 \cdot 40}{8} = 120$

48
Find the LCM of 56 and 21.

$\frac{56 \cdot 21}{7} = 168$

49
Find the LCM of 72 and 32.

288

50
Find the LCM of 18 and 23. 414

51
Find the LCM of 35 and 60. 420

FEEDBACK UNIT 7

This quiz reviews the preceding unit. Answers are at the back of the book.

1. Factors of 30 are 3 and ______.
2. Factors of 24 are 8 and ______.
3. Using the counting numbers, write all the multiplication expressions that have a product of 10.
4. Write the set of counting-number factors of 14.
5. Write the set of counting-number factors of 25.
6. Write the set of counting-number factors of 16.
7. Write the set of counting-number factors of 30.
8. Write the set of counting-number factors of 32.
9. Write the set of counting-number factors of 31.
10. What is the HCF of 12 and 30?
11. What is the HCF of 10 and 40?
12. What is the HCF of 7 and 12?
13. What is the HCF of 8 and 20?
14. What is the HCF of 18 and 42?
15. What is the LCM of 4 and 8?
16. What is the LCM of 12 and 16?
17. What is the LCM of 4 and 7?
18. What is the LCM of 12 and 30?
19. What is the LCM of 15 and 35?
20. What is the LCM of 14 and 21?

UNIT 8: APPLICATIONS

In this Applications Section, the format of the text has been altered. Answers for the problems appear beneath them rather than in the right-hand column. Your studying emphasis should be on learning the best procedures to follow with word problems. For that reason, once the procedure is learned a calculator may be used to complete the answer.

1

A rancher has three pastures measuring 4,352 acres, 518 acres, and 47 acres. Find the rancher's total pasture land.

Answer: The clue word in the problem is **total**. Since "total" is the answer to an addition problem, the word indicates the numbers should be added.

$$\begin{array}{r} 4{,}352 \\ 518 \\ +\ \ 47 \\ \hline 4{,}917 \end{array}$$

The total pasture land is 4,917 acres.

2

Paul is 18 years old and his father is 41. What is the difference in their ages?

Answer: The word **difference** is the clue to solving the problem because "difference" is the answer to a subtraction.

$$\begin{array}{r} 41 \\ -\ 18 \\ \hline 23 \end{array}$$

23 years is the difference in their ages.

3
Laura earns $52 each week. How many weeks does Laura need to work to earn $1,000?

Answer: The word **each** is a clue to solving the problem because the word usually requires multiplication or division. In this case 1,000 should be divided by 52.

$1000 \div 52 = 19 \text{ R}12$

The quotient (answer) has a remainder of 12.
This means that after 19 weeks Laura will need $12 more to reach her $1,000 goal. The answer to the problem is best stated as 20 weeks.

4
A produce store received 15 crates with 12 heads of cabbage in each crate. How many heads of cabbage were there?

Answer: The word **each** is a clue to solving the problem because the word usually requires multiplication or division. In this case 15 should be multiplied by 12.

$15 \cdot 12 = 180$

The product (answer) is 180 heads of cabbage.

5
Sam Jones owes two bills of $512 and $87.
Find the sum of money owed by Sam.

Answer: The word **sum** is the clue to solving the problem because it is the answer to an addition.

$$\begin{array}{r} 512 \\ +\ 87 \\ \hline 599 \end{array}$$

$599 is the sum of money owed by Sam.

6

If Sara has $15 and Mike has $38, how many more dollars does Mike have?

Answer: The phrase **how many more** is the clue to solving the problem because it indicates subtraction is needed.

$$\begin{array}{r} 38 \\ -\ 15 \\ \hline 23 \end{array}$$

Mike has $23 more than Sara.

7

Jane is a professional tutor and has 32 tutoring sessions each week. In a 16-week semester, how many tutoring sessions does she have?

Answer: The word **each** is a clue that the problem requires multiplication.

$32 \bullet 16 = 512$

Jane has 512 tutoring sessions in a 16-week semester.

8

On his math tests Mark has scored 95, 78, 68, and 91. What is his average test score?

Answer: The word **average** is the clue to solving the problem. To find an average, add the numbers and divide by the number of addends.

$$\begin{array}{r} 95 \\ 78 \\ 68 \\ +\ 91 \\ \hline 332 \end{array}$$

$332 \div 4 = 83$

Mark's average test score is 83.

9

A trapezoid has sides of 18 inches, 41 inches, 20 inches, and 47 inches. What is its perimeter?

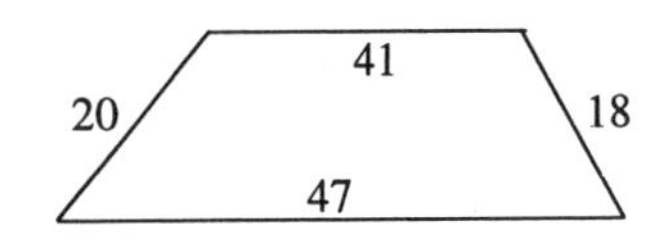

Answer: Often it is a good idea to draw a picture of the problem situation. A trapezoid is a four-sided figure with one pair of parallel sides. The word **perimeter** is the clue to solving the problem because it indicates the lengths should be added.

$$\begin{array}{r} 18 \\ 41 \\ 20 \\ +\ 47 \\ \hline 126 \end{array}$$

The perimeter of the trapezoid is 126 inches.

10

The Browns need to sod their back yard which is a rectangular shape that is 40 feet long and 23 feet wide. How many square feet of space needs to be covered?

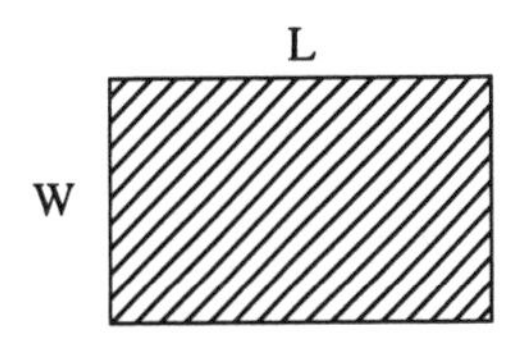

Answer: A drawing will help solve this problem. The term **square feet** is a clue to solving the problem because it indicates the **area** of the rectangle is needed. The formula for finding a rectangle's area is A = LW where A is the area in square units, L is the length in linear units, and W is the width in linear units.

$$\begin{aligned} A &= LW \\ &= 40 \bullet 23 \\ &= 920 \end{aligned}$$

The area to be sodded is 920 square feet.

FEEDBACK UNIT 8 FOR APPLICATIONS

1. **If Dan weighs 158 pounds and Joe weighs 141 pounds, how much more does Dan weigh?**
2. **Bill loaded the following amounts of concrete onto a truck: 100 lbs, 50 lbs, 25 lbs, 93 lbs. Find the total.**
3. **Find the perimeter of a triangle which has sides of length 34 inches, 43 inches, and 39 inches.**
4. **The Smiths are going to carpet their living room floor. The shape is a rectangle with length 17 feet and width 11 feet. How many square feet are in the living room floor?**
5. **If 16 employees have total wages of $5,712, find the average pay of each employee.**

SUMMARY FOR THE ARITHMETIC OF THE COUNTING NUMBERS

The following mathematical terms are crucial to an understanding of this chapter.

Set	Even numbers
Odd numbers	Between
Empty set	Element
Member	Counting numbers
Addition	Sum
Square brackets	Subtract
Difference	Multiplication
Product	Multiplication expression
Total	Division
Quotient	Numerical expression
Evaluating numerical expressions	Order of Operations
Factor	Highest common factor (HCF)
Multiple	Least common multiple (LCM)

Set is a mathematical word to describe a collection of persons, objects, or numbers. Braces, { }, are used to indicate a set. The symbols $\in$ and $\notin$ are used to make statements about members or elements of a set.

The arithmetic of the set of counting numbers was studied by learning to evaluate counting-number expressions for the different operations of arithmetic.

The operations of addition and multiplication were shown to be directly related to the counting process.

The relationship between the addition statement $8 + 5 = 13$ and the subtraction statement $13 - 5 = 8$ was shown. Similarly, the relationship between the multiplication statement $9 \cdot 5 = 45$ and the division statement $45 \div 5 = 9$ was shown.

Basic methods for determining the highest common factor (HCF) and least common multiple (LCM) of two counting numbers were also explained in Chapter 1. The highest common factor and least common multiple are necessary concepts to know when working with fractions later in the text.

CHAPTER 1 MASTERY TEST

The following questions test the objectives of Chapter 1. Answers are at the back of the book. The number in parentheses which follows each problem indicates the unit in which it can be learned.

1. Using braces, write the set of numbers 4, 7, and 15. (1)
2. Write the set of even numbers between 10 and 17. (1)
3. Write the set of odd numbers between 14 and 20. (1)
4. Write the set of counting numbers larger than 12 and less than 9. (1)
5. Write the set of even numbers between 28 and 30. (1)
6. $19 \in \{1, 2, 3, \ldots\}$. True or false? (1)
7. $2 \notin \{5, 7, 2, 8\}$. True or false? (1)
8. $26 \in \{2, 4, 6, \ldots\}$. True or false? (1)
9. $4 \notin \{\ \}$. True or false? (1)
10. $58 \in \{1, 2, 3, \ldots, 50\}$. True or false? (1)
11. Evaluate. $12 + 9$ (2)
12. Evaluate. $(9 + 6) + 4$ (2)
13. Evaluate. $(19 - 6) - 8$ (3)
14. Evaluate. $21 - (28 - 10)$ (3)
15. Evaluate. $(9 \bullet 3) \bullet 5$ (4)
16. Evaluate. $5 \bullet (3 \bullet 7)$ (4)
17. Evaluate. $(72 \div 4) \div 9$ (5)
18. Evaluate. $24 \div (6 \div 2)$ (5)
19. Evaluate. $3 \bullet 8 + 5$ (6)
20. Evaluate. $6 + 5 \bullet 3$ (6)
21. Evaluate. $[3 \bullet (4 + 2)] \bullet 2$ (6)
22. Evaluate. $14 + [(3 \bullet 2) + 12]$ (6)
23. Evaluate. $6 \bullet [3 + 6 \bullet 2]$ (6)
24. Evaluate. $[12 + 9 \bullet 3] + 13$ (6)
25. Write the set of counting-number factors of 20. (7)
26. Find the highest common factor (HCF) of 12 and 30. (7)
27. Find the highest common factor (HCF) of 9 and 15. (7)
28. Find the highest common factor (HCF) of 13 and 18. (7)
29. Find the least common multiple (LCM) of 6 and 9. (7)
30. Farmer A has 853 acres and Farmer B has 290 acres. What is the difference in the size of their farms? (8)
31. Find the sum of Julie Black's investments if she has separately invested $46,724 and $9,764. (8)

Chapter 2 Objectives

The following problems illustrate the objectives of this chapter. At this time you are not expected to know how to do these problems. However, if all of these problems are thoroughly understood, proceed directly to the Chapter 2 Mastery Test. The number in parentheses which follows each problem indicates the unit in which it can be learned.

1. Which of the following is/are numerical expressions? (1)
   a. $7 + 3 \cdot 41$
   b. $4 \cdot 7 + x$
   c. $9 \cdot 17 + 8 \cdot 12$
   d. $5(x + 47)$

2. Which of the following is/are open expressions? (1)
   a. $17 + 6x$
   b. $93 + 5 \cdot 13$
   c. $x + 98$
   d. $9(3x + 8)$

3. Evaluate $5 + 7x$ when $x = 3$. (2)

4. Evaluate $8(x + 3)$ when $x = 4$. (2)

5. Evaluate $2x + 4y$ when $x = 5$ and $y = 3$. (2)

6. Use the Commutative Law of Addition to write an equivalent expression for $7x + 19$. (3)

7. Use the Associative Law of Addition to write an equivalent expression for $(9x + 8) + 13$. (3)

8. Use the Commutative Law of Multiplication to write an equivalent expression for $x \cdot 7$. (4)

9. Use the Associative Law of Multiplication to write an equivalent expression for $8(9y)$. (4)

10. Use the Multiplication Law of One to write an equivalent expression for x. (4)

11. Use the Distributive Law of Multiplication over Addition to write an equivalent expression for $8x + 13x$. (5)

12. Write a simpler, equivalent expression for $5 + 8x + 7 + 9x$. (5)

13. Write a simpler, equivalent expression for $6(3x + 5) + 5(7x + 3)$. (6)

14. Write a simpler, equivalent expression for $(6x + 7) + 9(8x + 3)$. (6)

Chapter 2
The Algebra of the Counting Numbers

Unit 1: Numerical and Open Expressions

The following mathematical terms are crucial to an understanding of this unit.

Numerical expression
Open expression
Evaluate

1
❒ + 3 is not a numerical expression because the box is not a counting number. 5 + 3 is a **numerical expression** but ❒ + 5 ______ (is, is not) a numerical expression.

is not

2

❐ + 8 is not a numerical expression. It is an **open expression**. ❐ + 13 is not a numerical expression; it is an ______ ______.

open expression

3

❐ + 7 is an ______ expression.

open

4

9 + ❐ is an ______ ______.

open expression

5

3 + 5 is a numerical expression. 3 + ❐ is an open expression. 7 + ❐ is an ______ ______.

open expression

6

13 + ❐ is an ______ ______.

open expression

7

(3 + 4) + 2 is a numerical expression.
(5 + 6) + ❐ is an open expression.
(9 + 2) + ❐ is an ______ expression.

open

8

4 + 7 • (❐ + 2) is an ______ ______.

open expression

9

6 + 3 • (5 + 1) is a ______ expression.

numerical

10

(4 + ❐) • (3 + 7) is an ______ expression.

open

11

(8 + 2) • (5 + 3) is a ______ expression.

numerical

12

[8 + 3 • ❐] + 4 is an ______ ______.

open expression

13

3 + 5 • [(4 + 2) • 7] is a ______ ______.

numerical expression

14

The box in ❐ + 3 represents a place to put a counting number. What is supposed to be put in the place of the box in ❐ + 5?

a counting number

15
The box in 13 + ❐ represents a place to put a ______ ______.

counting number

16
The box in 5 + ❐ represents a place for a ______ number.

counting

17
If 3 is used to replace the box of ❐ + 17, the result is the numerical expression 3 + 17. If 7 is used to replace the box of ❐ + 17, the result is ________.

7 + 17, which is a numerical expression

18
Replacing the box of the open expression 8 + ❐ with a 5 results in the numerical expression 8 + 5. What is obtained by replacing the box of 10 + ❐ by 6?

10 + 6, which is a numerical expression

19
Whenever the box of 5 • ❐ + 2 is replaced by a counting number, the result is a numerical expression. By replacing the box of 5 • ❐ + 2 by 8, the result is ________ which is a __________ ____________.

5 • 8 + 2
numerical expression

20
Any counting number can replace the box of ❐ + 9. Replace the box by 4 and write the numerical expression.

4 + 9

21
Write the numerical expression obtained from the open expression 3 + ❐ by replacing the box by 6.

3 + 6

22
Any counting number may replace the box of 3 • ❐ + 2. The box of 3 • ❐ + 2 may be replaced by ______ number from the set {1, 2, 3, . . . }.

any

23
$21 \in \{1, 2, 3, \ldots\}$. Can 21 replace the box of 6 • ❐?

Yes

24
$8 \in \{1, 2, 3, \ldots\}$. Can 8 replace the box of 5 + ❐?

Yes

25
If the box of ❒ + 19 is replaced by 4, the result is 4 + 19, which has an evaluation of 23. If the box of ❒ + 13 is replaced by 9, the result is 9 + 13, which has an evaluation of ______.

22

26
Replace the box of ❒ + 8 by 3 and **evaluate** the numerical expression.

$3 + 8 = 11$

27
Replace the box of 3 • ❒ by 5 and evaluate the numerical expression.

$3 \cdot 5 = 15$

28
Replace the box of 8 • (❒ + 1) by 3 and evaluate.

$8 \cdot 4 = 32$

29
Replace the box of (❒ + 3) • 2 by 6 and evaluate.

$9 \cdot 2 = 18$

30
Replace the box of (1 + 2) • (❒ + 3) by 4 and evaluate.

$3 \cdot 7 = 21$

31
In the open expression ❒ • (2 + ❒) any number used as a replacement for one box must also be used to replace the other box. If 3 is used as a replacement for one box, what number must be used to replace the other box?

3

32
If 5 is used as the replacement for the boxes of (3 + ❒) • ❒, the numerical expression is (3 + 5) • 5. If 7 is used as the replacement in (3 + ❒) • ❒, the numerical expression is ______.

$(3 + 7) \cdot 7$

33
Replace both boxes of ❒ • (5 + ❒) by 4 and evaluate the numerical expression.

$4 \cdot (5 + 4) = 4 \cdot 9 = 36$

34
Replace the boxes of 8 • ❒ + ❒ by 9 and evaluate.

$8 \cdot 9 + 9 = 72 + 9 = 81$

35
Replace the boxes of ❒ • ❒ + 3 by 2 and evaluate.

$2 \cdot 2 + 3 = 4 + 3 = 7$

36
Evaluate ❒ + 10 when the box is replaced by 5. 15

37
Evaluate the open expression 2 • ❒ when the box is replaced by 7. 14

38
For the open expression 2 • ❒ + 5 any counting number may replace the box. Evaluate 2 • ❒ + 5 when the box is replaced by 8. 21

39
Evaluate the expression ❒ • 3 + 2 when the box is replaced by 6. 20

40
Evaluate (❒ + 3) • 5 when the box is replaced by 4. 35

41
Evaluate ❒ • (2 + ❒) when the box is replaced by 3. 15

42
Evaluate ❒ • ❒ + 3 when the box is replaced by 4. 19

43
Evaluate (2 + ❒) • 3 when the box is replaced by 1. 9

44
Evaluate 3 • [2 • ❒ + 4] when the box is replaced by 15. 102

45
Evaluate ❒ • [(2 + ❒) • 3] when the box is replaced by 2. 24

46
Evaluate ❒ • [(2 + ❒) • 3] when the box is replaced by 10. 360

47
Evaluate ❒ • [(2 + ❒) • 3] when the box is replaced by 4. 72

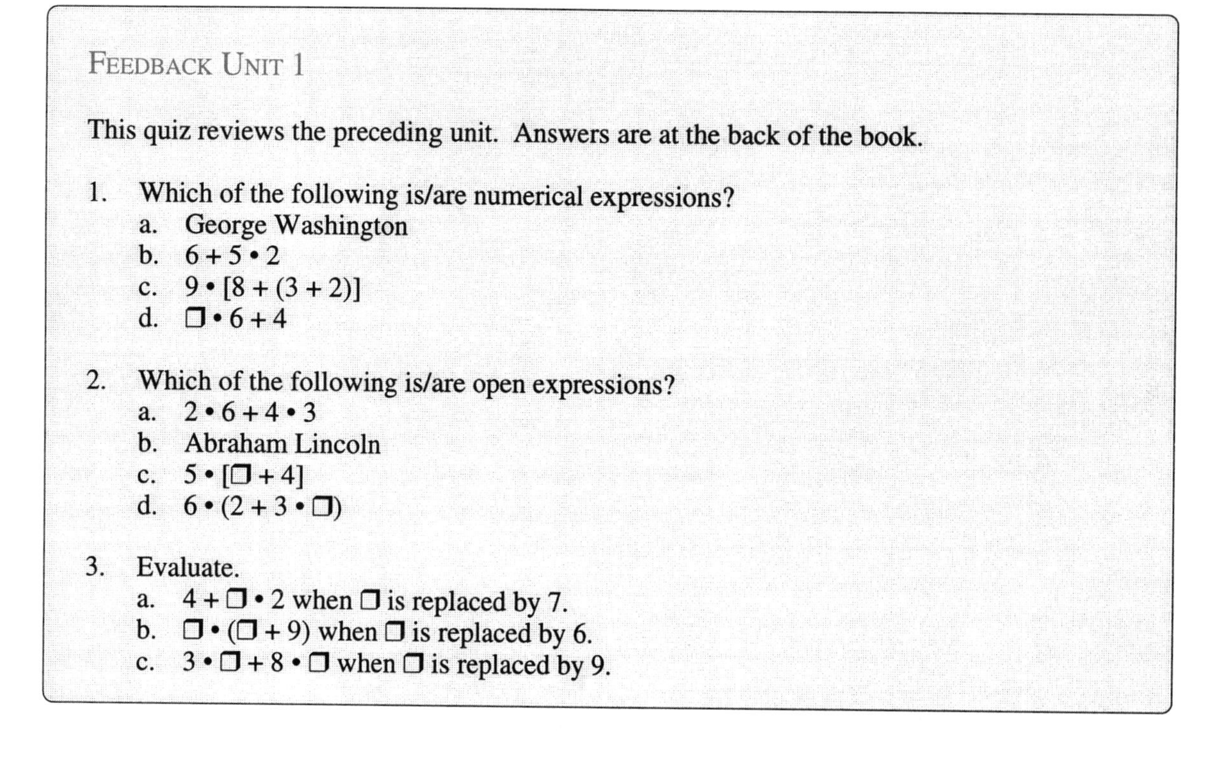

FEEDBACK UNIT 1

This quiz reviews the preceding unit. Answers are at the back of the book.

1. Which of the following is/are numerical expressions?
   a. George Washington
   b. 6 + 5 • 2
   c. 9 • [8 + (3 + 2)]
   d. □ • 6 + 4

2. Which of the following is/are open expressions?
   a. 2 • 6 + 4 • 3
   b. Abraham Lincoln
   c. 5 • [□ + 4]
   d. 6 • (2 + 3 • □)

3. Evaluate.
   a. 4 + □ • 2 when □ is replaced by 7.
   b. □ • (□ + 9) when □ is replaced by 6.
   c. 3 • □ + 8 • □ when □ is replaced by 9.

UNIT 2: EVALUATING NUMERICAL EXPRESSIONS

The following mathematical terms are crucial to an understanding of this unit.

Evaluating open expressions   Variable

1
Any **letter** of the alphabet can be used in place of the box in an open expression. The open expression □ + 9 can be written x + 9 or a + 9. Use the letter x to write the open expression □ + 3.

x + 3

2

$x + 4$ is the same **open expression** as ❒ + 4.
Use the letter k to write the open expression 5 + ❒.

$5 + k$

3

When a letter of the alphabet is used instead of a box in an open expression, the letter is called a **variable**. Use the letter b to write the open expression 7 + ❒.

$7 + b$

4

Use the variable m to write 4 • ❒ + 3.

$4 \bullet m + 3$

5

Use the variable z to write 16 + 5 • ❒.

$16 + 5 \bullet z$

6

Use the variable w to write (❒ + 5) • ❒.

$(w + 5) \bullet w$

7

❒ + 13 is the same as $x + 13$. Since the box may be replaced by any counting number, the variable x may be replaced by any ______ ______.

counting number

8

In the open expression $3 \bullet x + 17$, the variable x may be replaced by ______ counting number.

any

9

Multiplication is indicated whenever a variable and a number appear next to each other. $7 \bullet a$ can be written **without the dot** as 7a. Similarly, $4 \bullet a$ can be written without the dot as ______.

4a

10

$19 \bullet c$ can be written without the dot as ______.

19c

11

12 times x can be written $12 \bullet x$ or ______.

12x

12

5 times r can be written $5 \bullet r$ or ______.

5r

13

3x means to multiply 3 and x.
7x means to ______ 7 and x.

multiply

14

To **evaluate** 7t when t is replaced by 3, the numerical expression is 7 • 3. To evaluate 7t when t is replaced by 5, we obtain 7t = 7 • 5 = ______.

35

15

Evaluate 3b when b is replaced by 8.
[Note: 3b = 3 • b.]

24

16

Evaluate 5n when n is replaced by 6.

30

17

The open expression ❒ + 3 can be written as x + 3. Evaluate x + 3 when x is replaced by 5.

8

18

To evaluate the open expression 3x + 4, when x is replaced by 5, the following steps are used:

3x + 4
3 • 5 + 4
15 + 4
19

Evaluate 3x + 4 when x is replaced by 8.

28

19

Evaluate 3x + 4 when x is replaced by 10.

34

20

Evaluate 3x + 4 when x is replaced by 14.

46

21

Evaluate 5 + 2x when x is replaced by 4.

13

22

Evaluate 5 + 2x when x is replaced by 7.

19

23

Evaluate 5 + 2x when x is replaced by 10.

25

24

Evaluate 5 + 2x when x is replaced by 15.

35

25

Evaluate 5t + 3 when t is replaced by 7.

38

26
Evaluate $10y + 1$ when y is replaced by 5.

51

27
To evaluate the open expression $3(x + 5)$ when $x = 4$, the following steps are used:

$$3(x + 5)$$
$$3(4 + 5)$$
$$3 \cdot 9$$
$$27$$

Evaluate $3(x + 5)$ when $x = 2$.

21

28
Evaluate $3(x + 5)$ when $x = 10$.
($x = 10$ means x is to be replaced by 10.)

45

29
Evaluate $2x + 5x$ when $x = 7$.

$2 \cdot 7 + 5 \cdot 7 = 49$

30
Evaluate $2x + 5x$ when $x = 6$.

$12 + 30 = 42$

31
Evaluate $2x + 5x$ when $x = 10$.

70

32
Evaluate $(5 + x) + 3$ when $x = 1$.

$(5 + 1) + 3 = 6 + 3 = 9$

33
Evaluate $(5 + x) + 3$ when $x = 21$.

$26 + 3 = 29$

34
Evaluate $(5 + x) + 3$ when $x = 15$.

23

35
To evaluate $x(x +5)$ when $x = 2$, the following steps are used:

$$x(x + 5)$$
$$2(2 + 5)$$
$$2 \cdot 7$$
$$14$$

Note: $2(7)$ means $2 \cdot 7$

Evaluate $x(x + 5)$ when $x = 5$.

$5(10) = 5 \cdot 10 = 50$

36
Evaluate $x(x + 5)$ when $x = 3$.

24

37
Evaluate $x(x + 3) + x$ when $x = 7$.

77

38
In the open expression $(x + y) + 3$, there are two different variables. Since the variables are different, the numbers to replace them can be different. Write the **numerical expression** for $(x + y) + 3$ when $x = 2$ and $y = 5$.

$(2 + 5) + 3$

39
$(x + y) + 5$ when $x = 3$ and $y = 7$ is $(3 + 7) + 5$.
Write the numerical expression for $(x + y) + 10$ when $x = 5$ and $y = 3$.

$(5 + 3) + 10$

40
Write the numerical expression for $x + (y + 3)$ when $x = 7$ and $y = 5$.

$x + (y + 3)$
$7 + (5 + 3)$

41
Write the expression for $a + (8 + b)$ when $a = 2$ and $b = 7$.

$a + (8 + b)$
$2 + (8 + 7)$

42
To evaluate $d + (4 + e)$ when $d = 5$ and $e = 8$,

$$d + (4 + e)$$
$$5 + (4 + 8)$$
$$5 + 12$$
$$17$$

What is the evaluation of $d + (4 + e)$ when $d = 7$ and $e = 3$?

14

43
Evaluate the open expression $(m + 3) + n$ when $m = 6$ and $n = 10$.

19

44
Evaluate $k + (d + 3)$ when $k = 10$ and $d = 2$.

15

45
Evaluate $3 + (p + q)$ when $p = 9$ and $q = 6$.

18

46
Evaluate $(b + 7) + a$ when $a = 4$ and $b = 7$.

18

47
In the open expression $(x + y) + z$, there are three different variables. Each different variable may be replaced by any counting number. Write the numerical expression for $(x + y) + z$ when $x = 2$, $y = 5$, and $z = 7$.

$(x + y) + z$
$(2 + 5) + 7$

48
Write the open expression $a + (b + c)$ as a numerical expression when $a = 5$, $b = 8$, and $c = 10$.

$5 + (8 + 10)$

49
Evaluate $(p + q) + r$ when $p = 9$, $q = 7$, and $r = 10$.

26

50
Evaluate $(x + z) + t$ when $t = 3$, $x = 7$, and $z = 9$.

19

51
To evaluate $y + 3x$ when $x = 5$ and $y = 7$, the following steps are used:

$$y + 3x$$
$$7 + 3 \cdot 5$$
$$7 + 15$$
$$22$$

Evaluate $y + 3x$ when $x = 10$ and $y = 9$.

39

52
Evaluate $5x + y$ when $x = 6$ and $y = 7$.

37

53
Evaluate $z + 9b$ when $b = 5$ and $z = 6$.

51

54
Evaluate $x + 5y$ when $x = 7$ and $y = 3$.

22

55
Evaluate $2x + 5y$ when $x = 7$ and $y = 3$.

29

56
Evaluate $(7 + 3a) + b$ when $a = 5$ and $b = 2$.

24

57
Evaluate $3x + 3y$ when $x = 4$ and $y = 5$.

27

58
Evaluate $10m + n$ when $m = 3$ and $n = 1$.

31

59
Evaluate $2x + (3y + 5)$ when $x = 3$ and $y = 5$. 26

60
Evaluate $x(5 + 2y)$ when $x = 3$ and $y = 2$. 27

61
Evaluate $(5r + 3s) + 6t$ when $r = 3$, $s = 4$, and $t = 1$. 33

62
Evaluate $2x + (y + 3z)$ when $x = 4$, $y = 7$, and $z = 1$. 18

63
Evaluate $2(x + 3y) + 4z$ when $x = 4$, $y = 2$, and $z = 7$. 48

64
Evaluate $x + [2x + (y + 3)]$ when $x = 3$ and $y = 7$. 19

65
Evaluate $c(3a + 2b)$ when $a = 2$, $b = 4$, and $c = 5$. 70

66
Evaluate the open expression $ab + 7$ when $a = 3$ and $b = 4$.
[Note: $ab = a \bullet b$.] 19

67
Evaluate $rs + r$ when $r = 4$ and $s = 5$. 24

68
Evaluate $5x + [2a + (5 + b)]$ when $a = 3$, $b = 10$, and $x = 2$. 31

FEEDBACK UNIT 2

This quiz reviews the preceding unit. Answers are at the back of the book.

1. Evaluate $x + 5$ when $x = 3$.
2. Evaluate $2x + 4$ when $x = 7$.
3. Evaluate $5(x + 4)$ when $x = 3$.
4. Evaluate $3x$ when $x = 7$.
5. Evaluate $x(9 + x)$ when $x = 2$.
6. Evaluate $x + 3y$ when $x = 4$ and $y = 5$.
7. Evaluate $(x + z) + t$ when $t = 3$, $x = 7$, and $z = 1$.
8. Evaluate $a + (12 + b)$ when $a = 3$ and $b = 5$.
9. Evaluate $5 + xy$ when $x = 4$ and $y = 7$.
10. Evaluate $a(5 + 2b)$ when $a = 3$ and $b = 2$.

UNIT 3: COMMUTATIVE AND ASSOCIATIVE LAWS OF ADDITION

The following mathematical terms are crucial to an understanding of this unit.

Equivalent expressions	Commutative Law of Addition
Addends	Substitution
Associative Law of Addition	Simplifying open expressions

Two open expressions are **equivalent** whenever they have the same evaluation regardless of what number is used to replace the letter(s) or variable(s). In this section equivalent open expressions are explained.

1
Evaluate x + 8 when x = 11.

19

2
Evaluate (x + 3) + 5 and x + 8 when x = 7. Are the evaluations of the two expressions the same?

Yes, both are 15

3
Evaluate the two open expressions (2 + y) + 9 and 11 + y when y = 3. Are the evaluations the same?

Yes, both are 14

4
Evaluate the two open expressions (6 + a) + 7 and 13 + a when a = 3. Are the evaluations the same?

Yes, both are 16

5
Evaluate the two open expressions 8x + 3x and 11x when x = 3. Are the evaluations the same?

Yes, both are 33

6
Evaluate 3x + 4 and 7x when x = 2.
Are the evaluations the same?

No, 3x + 4 is 10
and 7x is 14

7
Evaluate $(2x + 3x) + 6$ and $5x + 6$ when $x = 3$.
Are the evaluations the same?

Yes, both are 21

8
Evaluate $2(x + 3)$ and $2x + 6$ when $x = 7$.
Are the evaluations the same?

Yes, both are 20

9
Evaluate $x + y$ and $y + x$ when $x = 2$ and $y = 5$.
Are the evaluations the same?

Yes, both are 7

10
$x + y$ and $y + x$ have the same evaluations for any number replacements of x and y which means that $x + y$ and $y + x$ are **equivalent expressions**. Are $z + w$ and $w + z$ equivalent expressions?

Yes. Choose any numbers for z, w. Evaluations are equal.

11
$r + 4$ and $4 + r$ are equivalent expressions because they have the same evaluation for any number replacement of r. Are $x + 9$ and $9 + x$ equivalent expressions?

Yes. Choose any number for x. Evaluations are equal.

12
The fact that $x + y$ and $y + x$ are equivalent expressions is called the **Commutative Law of Addition.** According to the Commutative Law of Addition, $x + 34$ is equivalent to ______.

$34 + x$

13
The Commutative Law of Addition allows $3 + 18$ to be replaced by $18 + 3$. The **order of two addends** can be reversed without changing the sum. Write $3 + 18$ with the order of the two addends reversed.

$18 + 3$

14
$x + 5$ is equivalent to $5 + x$ by the Commutative Law of Addition which allows the order of two addends to be reversed. Write $k + z$ with the order of the two addends reversed.

$z + k$

15
The Commutative Law of Addition states that $a + 3 =$ ______. | $3 + a$

16
The expression $37 + x$ may be **substituted** for ______ by the Commutative Law of Addition. | $x + 37$

17
$11 + y$ may be substituted for ______ by the Commutative Law of Addition. | $y + 11$

18
Change $a + 8$ to an equivalent expression by using the Commutative Law of Addition. | $8 + a$

19
$2x + 7$ is equivalent to $7 + 2x$ because of the Commutative Law of Addition. Use the Commutative Law of Addition to write an equivalent expression for $5a + 3$. | $3 + 5a$

20
$17x + 3$ may be substituted for ______ by the Commutative Law of Addition. | $3 + 17x$

21
$3a + 5b$ may be substituted for ______ by the Commutative Law of Addition. | $5b + 3a$

22
In the addition of two counting numbers the order of the addends can be reversed by the Commutative Law of Addition. $3x + 9a$ is equivalent to ______. | $9a + 3x$

In the previous frames it has been shown that $x + y$ is equivalent to $y + x$ by the Commutative Law of Addition. This means that the order of addends may be reversed without changing their sum.

The following frames deal with the addition of three numbers. Two equivalent expressions will be shown for the sum of any three counting numbers.

23

$(9 + x) + 3$ and $9 + (x + 3)$ have the same evaluations when $x = 4$. Do the open expressions have the same evaluations when $x = 15$?

Yes, both are 27

24

$(9 + x) + 3$ and $9 + (x + 3)$ have the same evaluations for any number replacement of x. Are the expressions equivalent?

Yes

25

The open expressions $(x + y) + z$ and $x + (y + z)$ have the same evaluations for any replacements of x, y, and z. Are the expressions equivalent?

Yes

26

The fact that $(x + y) + z$ and $x + (y + z)$ are equivalent expressions is the **Associative Law of Addition**. According to the Associative Law of Addition, $(16 + y) + 3$ is equivalent to ______.

$16 + (y + 3)$

27

$(a + b) + c$ and $a + (b + c)$ are equivalent expressions. Are $(x + 5) + 7$ and $x + (5 + 7)$ equivalent expressions?

Yes

28

$(x + y) + z$ is equivalent to $x + (y + z)$. This is a statement of the Associative Law of Addition. According to the Associative Law of Addition, $(5 + y) + 3$ is equivalent to ______.

$5 + (y + 3)$

29

$(x + 3) + b$ is equivalent to $x + (3 + b)$ because of the Associative Law of Addition. $(x + 7) + w$ is equivalent to ______ by the Associative Law of Addition.

$x + (7 + w)$

30

$(8 + 17) + 3$ is equivalent to $8 + (17 + 3)$ by the ______ Law of Addition.

Associative

31
When the addition of three numbers is indicated by an example such as $4 + 7 + 3$, the Associative Law of Addition states that the sum obtained is the same whether the numbers are grouped $(4 + 7) + 3$ or ______.

$4 + (7 + 3)$

32
$(3 + 9) + 7$ is equivalent to $3 + (9 + 7)$.
When the numbers $5 + 8 + 10$ are grouped by the Associative Law of Addition, $(5 + 8) + 10$ is equivalent to ______.

$5 + (8 + 10)$

33
The expression $(5 + 3) + 4$ may be substituted for $5 + (3 + 4)$ by using the ______ Law of Addition.

Associative

34
The Associative Law of Addition states that $(x + 7) + p$ is equivalent to ______.

$x + (7 + p)$

35
By the Associative Law of Addition, $4 + (7 + 3)$ is equivalent to ______.

$(4 + 7) + 3$

36
By the Associative Law of Addition, $8 + (5 + x)$ is equivalent to ______.

$(8 + 5) + x$

37
By the Associative Law of Addition, $19 + (5 + t)$ is equivalent to ______.

$(19 + 5) + t$

38
By the Associative Law of Addition, $x + (9 + 7)$ is equivalent to ______.

$(x + 9) + 7$

In the preceding frames it has been shown that $(x + y) + z$ is equivalent to $x + (y + z)$ by the Associative Law of Addition. This means that the **grouping of three addends** may be changed without altering their sum.

39

The Commutative Law of Addition states that the order of two addends can be reversed. The Associative Law of Addition states that the **grouping of three addends** can be changed. The Associative Law of Addition states that $(4 + 9) + 3$ is equivalent to ______.

$4 + (9 + 3)$

40

$5x + (7 + 4)$ may be substituted for ______ by the Associative Law of Addition.

$(5x + 7) + 4$

41

$(7 + 3) + 9b$ may be substituted for ______ by the Associative Law of Addition.

$7 + (3 + 9b)$

42

$x + 7$ may be substituted for ______ by the Commutative Law of Addition.

$7 + x$

43

$(7x + 2x) + 9$ may be substituted for ______ by the Associative Law of Addition.

$7x + (2x + 9)$

44

When the order of two addends is reversed, it is a commutative property. When the grouping of three addends is changed, it is an associative property. Which property states that $(9 + r) + 6$ is equivalent to $(r + 9) + 6$?

Commutative
$9 + r = r + 9$

45

$(8 + x) + 3$ is equivalent to $(x + 8) + 3$ by the __________ Law of Addition.

Commutative
$8 + x = x + 8$

46

$4a + (9 + a)$ is equivalent to $(4a + 9) + a$ by the __________ Law of Addition.

Associative (Grouping has been changed.)

The Commutative (order) and Associative (grouping) Laws of Addition are used to write equivalent expressions. These laws are the basis for simplifying open expressions.

47
To **simplify** $3 + x + 5$, the order and grouping of the addends is changed. The simplification of $3 + x + 5$ is shown below:

$$3 + x + 5$$
$$(3 + x) + 5$$
$$(x + 3) + 5$$
$$x + (3 + 5)$$
$$x + 8$$

Is $3 + x + 5$ equivalent to $x + 8$?

Yes

48
To simplify $9 + z + 7$, the order and grouping of the addends is changed to write an equivalent expression.

$$9 + z + 7$$
$$z + (9 + 7)$$
$$z + 16$$

Is $9 + z + 7$ equivalent to $z + 16$?

Yes

49
The simplest equivalent expression for $6 + y + 2$ is $y + 8$. What is the simplest equivalent expression for $7 + k + 5$?

$k + 12$

50
To simplify $4 + x + 5$, change the ordering and grouping so that 4 and 5 are added.
Simplify $4 + x + 5$.

$4 + x + 5$
$x + (4 + 5)$
$x + 9$

51
Find the simplest equivalent expression for $7 + z + 16$.

$z + (7 + 16) = z + 23$

52
Simplify. $9 + a + 3$

$a + (9 + 3) = a + 12$

53
Simplify. $9 + r + 3$

$r + 12$

54
Simplify. $a + 4 + 9$

$a + (4 + 9)$
$a + 13$

55
Simplify. $6 + a + 11$

$a + 17$

56
Simplify. $7 + b + 15$

$b + 22$

57
Simplify. $x + 12 + 6$

$x + 18$

58
Simplify. $4 + x + 7$

$x + 11$

59
Simplify. $9 + g + 8$

$g + 17$

60
Simplify. $5 + r + 17$

$r + 22$

61
Simplify. $d + 8 + 5$

$d + 13$

62
Simplify. $8 + c + 4$

$c + 12$

63
Simplify. $23 + 4b + 15$

$4b + 38$

64
Simplify. $8 + 3h + 16$

$3h + 24$

FEEDBACK UNIT 3

This quiz reviews the preceding unit. Answers are at the back of the book..

1. For each of the following write an equivalent expression using the Commutative Law of Addition.
   a. $x + y =$ ______
   b. $13 + a =$ ______
   c. $w + 8 =$ ______
   d. $2k + 7 =$ ______

2. For each of the following write an equivalent expression using the Associative Law of Addition.
   a. $(x + y) + z =$ ______
   b. $(5 + x) + 9 =$ ______
   c. $(3x + 6) + 7 =$ ______
   d. $5 + (13 + 6x) =$ ______

3. Find a simpler equivalent expression for $(5x + 9) + 30$.

4. Find a simpler equivalent expression for $(6 + 7x) + 13$.

5. Find a simpler equivalent expression for $4 + (8x + 19)$.

6. Simplify. $x + 9 + 7$

7. Simplify. $4 + x + 8$

8. Simplify. $3 + 5 + y$

9. Simplify. $12 + x + 7$

10. Simplify. $4 + 3x + 7$

UNIT 4: COMMUTATIVE AND ASSOCIATIVE LAWS OF MULTIPLICATION AND THE MULTIPLICATION LAW OF ONE

The following mathematical terms are crucial to an understanding of this unit.

Commutative Law of Multiplication	Factors
Associative Law of Multiplication	Multiplication Law of One

1
Replace x by 5 and y by 3 in the open expressions xy and yx. Do the numerical expressions have the same evaluation?
[Note: xy = x • y.]

Yes, both are 15

2
Replace x by 6 and y by 8 in the open expressions xy and yx. Do the numerical expressions have the same evaluation?

Yes, both are 48

3
For any counting-number replacements for x and y in the open expressions xy and yx, will the evaluations be the same?

Yes

4
The fact that xy is equivalent to yx is the **Commutative Law of Multiplication**. According to the Commutative Law of Multiplication, 5 • 8 is equivalent to ______.

8 • 5

5
mn is **equivalent** to nm because of the Commutative Law of Multiplication. rs is equivalent to sr because of the ______ Law of Multiplication.

Commutative

6
By the Commutative Law of Multiplication, x • 8 is equivalent to ______.

8x (8x means 8 • x)

7
The expression 15 • 3 may be substituted for 3 • 15 by the ______ Law of Multiplication.

Commutative

8
The **order of two factors** may be reversed by the Commutative Law of Multiplication. By the Commutative Law of Multiplication, 9 • 3 has the same evaluation as ______.

3 • 9

9
The expression w • 7 may be replaced by ______ by the Commutative Law of Multiplication.

7w

10
The expression z • 17 may be replaced by ______ by the Commutative Law of Multiplication.

17z

11
The expression t • 46 may be replaced by ______ by the Commutative Law of Multiplication.

46t

12
Is (5 • 7) • 2 equal to 5 • (7 • 2)?

Yes, both are 70

13
If y is replaced by 8 and z is replaced by 2, do the open expressions (5y)z and 5(yz) have the same evaluation?

Yes, both are 80

14
If x = 2, y = 3, and z = 5, do the expressions (xy)z and x(yz) have the same evaluation?

Yes, both are 30

15
For any counting-number replacements for x, y, and z, do the open expressions (xy)z and x(yz) have the same evaluation?

Yes

16
The fact that (ab)c is equivalent to a(bc) is the **Associative Law of Multiplication**. By the Associative Law of Multiplication, 3 • (5 • 7) is equivalent to ______.

(3 • 5) • 7

17
3 • (5x) is equivalent to (3 • 5)x by the Associative Law of Multiplication. 7 • (4z) is equivalent to (7 • 4)z by the ______ Law of Multiplication.

Associative

18
7 • (6z) is equivalent to (7 • 6)z by the ______ Law of Multiplication.

Associative

19
Multiplication can only be performed on two numbers at a time. The Associative Law of Multiplication states that the **grouping of three factors** can be changed. (8 • 3) • 5 is equivalent to ______.

8 • (3 • 5)

20
The Commutative Law of Multiplication states that the order of two factors can be reversed. The Associative Law of Multiplication states that the grouping of three factors can be changed. The Associative Law of Multiplication states that (4 • 9) • 3 is equivalent to ______.

4 • (9 • 3)

21
3 • (5x) is equivalent to (3 • 5)x by the Associative Law of Multiplication because the grouping has been changed. 6 • (3x) is equivalent to ______ by the Associative Law of Multiplication.

(6 • 3)x

22
To **simplify** 8 • (4x), the following steps are used:

8 • (4x)
(8 • 4)x
32x

Therefore, 8 • (4x) is simplified to 32x.
Simplify. 7 • (3y)

(7 • 3)y
21y

23
Simplify. $9 \bullet (5x)$

$(9 \bullet 5)x$
$45x$

24
Simplify. $9 \bullet (4a)$

$(9 \bullet 4)a$
$36a$

25
Simplify. $6 \bullet (8x)$

$48x$

26
To simplify the expression $(3r) \bullet 5$, the following steps are used:

$(3r) \bullet 5$
$5 \bullet (3r)$
$(5 \bullet 3)r$
$15r$

Simplify. $(4r) \bullet 6$

$24r$

27
Simplify. $(5x) \bullet 9$

$45x$

28
Simplify. $(7x) \bullet 6$

$42x$

29
Simplify. $3 \bullet (5x)$

$15x$

30
Simplify. $(9a) \bullet 7$

$63a$

31
Simplify. $20 \bullet (3x)$

$60x$

32
Is $1 \bullet 7 = 7$ a true statement?

Yes

33
Is 1x equivalent to x if x is replaced by 5?
[Note: $1x = 1 \bullet x$]

Yes, both are 5

34
$1 \bullet 8 = 8$ and $1 \bullet 17 = 17$ are true equalities.
Is 1x equivalent to x?

Yes

35
The fact that 1x is equivalent to x is the **Multiplication Law of One**. $1 \cdot 96$ is equivalent to ______. 96

36
By the Multiplication Law of One, 5 is equivalent to $1 \cdot 5$. Similarly, by the Multiplication Law of One, 19 is equivalent to ______. $1 \cdot 19$

37
By the Multiplication Law of One, $1 \cdot (18 + 3x)$ is equivalent to $18 + 3x$. Similarly, $1 \cdot (2x + 3)$ is equivalent to ______. $2x + 3$

38
By the Multiplication Law of One, $5x + 7$ is equivalent to $1 \cdot (5x + 7)$. Similarly, $6x + 4$ is equivalent to ______. $1 \cdot (6x + 4)$

FEEDBACK UNIT 4

This quiz reviews the preceding unit. Answers are at the back of the book.

1. Use the Commutative Law of Multiplication to write an equivalent expression for each of the following.
   a. $x \cdot 18 =$ ______ b. $(6x) \cdot 9 =$ ______
2. Use the Associative Law of Multiplication to write an equivalent expression for each of the following.
   a. $6 \cdot (3x) =$ ______ b. $(4 \cdot x) \cdot 5 =$ ______
3. Use the Multiplication Law of One to write an equivalent expression for each of the following.
   a. $1 \cdot k =$ ______ b. $m =$ ______
   c. $1 \cdot (3x + 2) =$ ______ d. $1 \cdot (14x) =$ ______
4. Use the Associative Law of Multiplication to write a simpler equivalent expression for $7(5x)$.
5. Use the Multiplication Law of One to write a simpler equivalent expression for $1 \cdot (8x + 31)$.
6. Simplify. $5 \cdot (3x)$
7. Simplify. $(4x) \cdot 3$
8. Simplify. $2 \cdot (5x)$
9. Simplify. $7(3x)$
10. Simplify. $(8x) \cdot 4$

UNIT 5: THE DISTRIBUTIVE LAW AND SIMPLIFYING OPEN EXPRESSIONS WITH LIKE TERMS

The following mathematical terms are crucial to an understanding of this unit.

Distributive Law of Multiplication over Addition
Like terms

A number next to a parentheses means multiply. $5(4 + 7)$ means $5 \cdot (4 + 7)$ and $(2 + 6)4$ means $(2 + 6) \cdot 4$. The multiplication dot may be removed whenever a number is multiplied by an expression that is in a parentheses.

1
Do $8 \cdot 2 + 3 \cdot 2$ and $(8 + 3)2$ have the same evaluation?
[Note: $(8 + 3)2 = (8 + 3) \cdot 2$.] — Yes, both are 22

2
Do $7 \cdot 6 + 9 \cdot 6$ and $(7 + 9)6$ have the same evaluation? — Yes, both are 96

3
Do $3 \cdot 4 + 3 \cdot 6$ and $3(4 + 6)$ have the same evaluation? — Yes, both are 30

4
Do $5(3 + 8)$ and $5 \cdot 3 + 5 \cdot 8$ have the same evaluation? — Yes, 55

5
Do $5x + 4x$ and $(5 + 4)x$ have the same evaluation when $x = 7$? — Yes, 63

6
Do $9x + 2x$ and $(9 + 2)x$ have the same evaluation when $x = 5$? — Yes, 55

7
Do $5(y + 2)$ and $5y + 5 \cdot 2$ have the same evaluation when $y = 6$? — Yes, 40

8
Do a(b + c) and ab + ac have the same evaluation when a = 2, b = 5, and c = 6?

Yes, both are 22

9
Do a(b + c) and ab + ac have the same evaluation when a = 3, b = 5, and c = 2?

Yes, both are 21

10
Do a(b + c) and ab + ac have the same evaluation when a = 1, b = 4, and c = 2?

Yes, both are 6

11
Are a(b + c) and ab + ac **equivalent expressions**?

Yes

12
The fact that a(b + c) and ab + ac are equivalent is the **Distributive Law of Multiplication over Addition**. By the Distributive Law of Multiplication over Addition 8(5 + 4) is equivalent to ______.

8 • 5 + 8 • 4

13
To **simplify** 8x + 3x, the following steps are used:

8x + 3x
(8 + 3)x
11x

Therefore, 8x + 3x is simplified to 11x.
Simplify. 7x + 5x

(7 + 5)x
12x

14
Simplify 9z + 6z by using the Distributive Law of Multiplication over Addition.

(9 + 6)z
15z

15
Simplify. 12a + 7a

(12 +7)a
19a

16
Simplify. 18x + 3x

21x

17

To simplify $7x + x$, first replace x by 1x.

$$7x + x$$
$$7x + 1x$$
$$(7 + 1)x$$
$$8x$$

Simplify. $3y + y$

$3y + 1y$
$(3 + 1)y$
$4y$

18

Simplify. $5a + a$

$5a + 1a$
$(5 + 1)a$
$6a$

19

Simplify. $4x + x$

$(4 + 1)x$
$5x$

20

Simplify. $r + 12r$

$13r$

21

Simplify. $19x + 4x$

$23x$

22

Simplify. $x + x$

$1x + 1x$
$(1 + 1)x$
$2x$

23

Simplify. $3m + 9m$

$12m$

In the expression $3x + 5 + 7x + 9$ the 3x and 7x are **like terms.** 3x and 7x have the same variable (letter), x. Also 5 and 9 are **like terms**. Two addends are like terms when they have the same variable factors.

To simplify $3x + 5 + 7x + 9$, the like terms are grouped as shown below:

$$3x + 5 + 7x + 9$$
$$(3x + 7x) + (5 + 9)$$
$$10x + 14$$

$10x + 14$ cannot be simplified to 24 or 24x, because 10x and 14 are not like terms.

24

To simplify $5 + 8x + 2x$, the like terms are grouped:

$$5 + 8x + 2x$$
$$5 + (8x + 2x)$$
$$5 + 10x$$

$5 + 10x$ cannot be simplified further.

Simplify. $2 + 6x + 3x$

$9x + 2$ or $2 + 9x$
($11x$ is incorrect)

25

Simplify $2a + 7 + 6a$ by first grouping the like terms.

$8a + 7$ or $7 + 8a$
($15a$ is incorrect)

26

Simplify $5x + 3x + 7$ by first grouping the like terms.

$(5x + 3x) + 7$
$8x + 7$

27

To simplify $3y + 5 + 4y$, the following steps are used:

$$3y + 5 + 4y$$
$$(3y + 4y) + 5$$
$$7y + 5$$

Simplify. $7r + 4 + 5r$

$12r + 4$ ($16r$ is incorrect)

28

To simplify $3x + 5x + 7x$, the following steps may be used:

$3x + 5x + 7x$	or	$3x + 5x + 7x$
$3x + (5x + 7x)$		$(3x + 5x) + 7x$
$3x + 12x$		$8x + 7x$
$15x$		$15x$

Simplify. $4b + 6b + 3b$

$13b$

29

Simplify. $17a + 6a + 2a$

$25a$

30

Simplify. $3m + 8m + 6m$

$17m$

31

Simplify. $12x + 8x + 3x$

$23x$

32

Simplify. $3a + 2a + 7$

$5a + 7$

33

Simplify. $5y + 6 + 4y$

$9y + 6$

34
Simplify. $18t + 3t + 6t$

$27t$

35
Simplify. $5 + 7x + 9x$

$16x + 5$

36
To simplify $7 + 5x + 3$, add the counting numbers 7 and 3. The simplest equivalent expression for $7 + 5x + 3$ is $5x + 10$. Simplify. $4 + 9x + 8$

$9x + 12$ or $12 + 9x$
($21x$ is incorrect because $9x$ and 12 are not like terms)

37
Simplify. $8 + 3x + 5$

$3x + 13$
($16x$ is incorrect)

38
Simplify. $8 + 12 + 9x$

$20 + 9x$ or $9x + 20$
($29x$ is incorrect)

39
Simplify. $5x + 7 + 12$

$5x + 19$

40
Simplify. $9t + 7 + 4t$

$13t + 7$

41
To simplify $5x + 7 + 4x$, add $5x$ and $4x$. The simplest expression equivalent to $5x + 7 + 4x$ is $9x + 7$. Simplify. $8x + 4 + 2x$

$10x + 4$ or $4 + 10x$
($14x$ is incorrect)

42
Simplify. $9x + 3 + 7x$

$16x + 3$

43
Simplify. $5x + 7x + 15$

$12x + 15$

44
Simplify. $17 + 3x + 5x$

$8x + 17$

45
Simplify. $4x + 7 + x$
[Note: $x = 1x$ by the Multiplication Law of One.]

$5x + 7$

46

Simplify. $x + 8 + x$ — $2x + 8$

47

To simplify $5x + 3 + 6x + 7$,

$$5x + 3 + 6x + 7$$
$$(5x + 6x) + (3 + 7)$$
$$11x + 10$$

Simplify. $8x + 9 + 5x + 3$ — $13x + 12$

48

Simplify. $4x + 6 + 8x + 9$ — $12x + 15$

49

Simplify. $8x + 4 + 3x + 9$, by adding like terms. — $11x + 13$

50

Simplify. $9 + 3x + 2 + 5x$ — $8x + 11$

51

Simplify. $2x + 7 + 3x + 2$ — $5x + 9$

52

Simplify. $4 + 8x + x + 3$ — $9x + 7$

53

Simplify. $5x + 7 + 9 + x$ — $6x + 16$

54

Simplify. $3x + 2 + 4x + 5 + 6x$ — $13x + 7$

FEEDBACK UNIT 5

This quiz reviews the preceding unit. Answers are at the back of the book.

Simplify.

1. $6 + 9x + 4x$
2. $8x + 7 + 6x$
3. $9 + 7x + 13x$
4. $6x + 9 + x$
5. $5 + 4x + 3x$
6. $10 + 4x + 3 + 5x$
7. $8x + 13 + x + 5$
8. $5x + 3 + 2x + 4 + 6x$

UNIT 6: SIMPLIFYING OPEN EXPRESSIONS

The following mathematical terms are crucial to an understanding of this unit.

Removing parentheses

1

To **remove the parentheses** from $5(2x + 3)$,
5 is multiplied by 2x and 5 is also multiplied by 3.

$$5(2x + 3)$$
$$5 \bullet 2x + 5 \bullet 3$$
$$10x + 15$$

Remove the parentheses from $4(3x + 7)$ by multiplying 4 by 3x and also multiplying 4 by 7.

$12x + 28$

2

The parentheses of $8(x + 3)$ are removed by multiplying both x and 3 by 8.

$$8(x + 3)$$
$$8 \bullet x + 8 \bullet 3$$
$$8x + 24$$

Remove the parentheses from $6(x + 9)$.

$6x + 54$

3

To remove the parentheses from $7(x + 4)$,

$$7(x + 4)$$
$$7x + 7 \bullet 4$$
$$7x + 28$$

Remove the parentheses from $3(y + 5)$.

$3y + 3 \bullet 5$
$3y + 15$

4

To remove the parentheses in the expression $4(x + 2)$,

$$4(x + 2)$$
$$4x + 4 \bullet 2$$
$$4x + 8$$

Remove the parentheses. $7(x + 3)$

$7x + 21$

5

Remove the parentheses. $6(m + 5)$

$6m + 30$

6
Remove the parentheses. $9(2 + e)$ — $18 + 9e$

7
Remove the parentheses. $6(7 + r)$ — $42 + 6r$

8
Remove the parentheses. $3(z + 6)$ — $3z + 18$

9
Remove the parentheses. $7(2 + y)$ — $14 + 7y$

10
Remove the parentheses. $10(m + 2)$ — $10m + 20$

11
Remove the parentheses. $7(y + 9)$ — $7y + 63$

12
Parentheses can be removed by multiplication.
For example, to remove the parentheses from $(7x + 5)$,

$$(7x + 5)$$
$$1(7x + 5)$$
$$1 \bullet 7x + 1 \bullet 5$$
$$7x + 5$$

Remove the parentheses from $(4x + 3)$. — $1(4x + 3) = 4x + 3$

13
To remove the parentheses from $(6x + 9)$ a multiplier of 1 can be used. Remove the parentheses from $(6x + 9)$. — $6x + 9$

14
Whenever parentheses have no number shown as the multiplier, the number 1 can be used. Remove the parentheses from $(2x + 5)$. — $2x + 5$

15
Remove the parentheses. $(x + 7)$ — $x + 7$

16
Remove the parentheses. $(2x + 3)$ — $2x + 3$

17
Remove the parentheses. $3(1 + c)$ — $3 + 3c$

18
Remove the parentheses. $6(6 + x)$

$36 + 6x$

19
Remove the parentheses. $8(3 + a)$

$24 + 8a$

20
Remove the parentheses. $(5 + y)$

$5 + y$

21
To remove the parentheses of the expression $3(2x + 5)$,

$$3(2x + 5)$$
$$3 \cdot 2x + 3 \cdot 5$$
$$6x + 15$$

Remove the parentheses. $5(4x + 3)$

$20x + 15$

22
Remove the parentheses. $5(7x + 4)$

$35x + 20$

23
Remove the parentheses. $7(3x + 6)$

$21x + 42$

24
Remove the parentheses. $(2x + 7)$

$2x + 7$

25
To **simplify** $5 + 3(2x + 1)$, the parentheses are removed first.

$$5 + 3(2x + 1)$$
$$5 + 3 \cdot 2x + 3 \cdot 1$$
$$5 + 6x + 3$$
$$6x + 8$$

Remove the parentheses from $7 + 2(3x + 5)$ and complete the simplification.

$7 + 6x + 10$
$6x + 17$

26
To simplify $6 + 4(3x + 5)$, the parentheses are removed first.

$$6 + 4(3x + 5)$$
$$6 + 12x + 20$$
$$12x + 26$$

Remove the parentheses from $9 + 6(x + 7)$ and complete the simplification.

$9 + 6x + 42$
$6x + 51$

27

The parentheses of $7x + (4x + 9)$ are preceded by a plus sign. A multiplier of 1 can be used to remove the parentheses.

$$7x + (4x + 9)$$

$$7x + 1(4x + 9)$$

$$7x + 4x + 9$$

$$11x + 9$$

Remove the parentheses from $6x + (8x + 5)$ and complete the simplification.

$6x + 8x + 5$
$14x + 5$

28

To simplify $3 + 6(2x + 7)$, the parentheses are removed first.

$$3 + 6(2x + 7)$$

$$3 + 12x + 42$$

$$12x + 45$$

Simplify. $7x + 4(3x + 5)$

$19x + 20$

29

Simplify. $4 + 7(2x + 5)$

$14x + 39$

30

Simplify. $6x + 5(x + 3)$

$11x + 15$

31

Simplify. $8 + (3x + 2)$

$3x + 10$

32

Simplify. $7x + (3x + 5)$

$10x + 5$

33

Simplify. $9 + 5(x + 6)$

$5x + 39$

34

To simplify the expression $5x + 3(2x + 4) + 6$,

$$5x + 3(2x + 4) + 6$$

$$5x + 6x + 12 + 6$$

$$11x + 18$$

Simplify. $2x + 5(3x + 2) + 3$

$17x + 13$

35
To simplify $3x + 5(x + 4) + 7$,

$$3x + 5(x + 4) + 7$$
$$3x + 5x + 20 + 7$$
$$8x + 27$$

Simplify. $12x + 3(7 + 2x) + 9$ — $18x + 30$

36
Simplify. $9 + 2(9x + 1) + 3x$ — $21x + 11$

37
Simplify. $7 + (2x + 3) + 5x$ — $7x + 10$

38
Simplify. $3 + 7(2x + 4) + 3x$ — $17x + 31$

39
Simplify. $x + 2(3x + 1) + 7$ — $7x + 9$

40
To simplify $3(2x + 5) + 2(5x + 1)$,
first remove both parentheses.

$$3(2x + 5) + 2(5x + 1)$$
$$6x + 15 + 10x + 2$$
$$16x + 17$$

Simplify. $5(4x + 1) + 4(6x + 8)$ — $44x + 37$

41
Simplify. $7(2x + 4) + 3(x + 4)$ — $17x + 40$

42
Simplify. $4(3x + 2) + 2(x + 9)$ — $14x + 26$

43
Simplify. $6(x + 4) + (3x + 5)$ — $9x + 29$

44
Simplify. $4(3x + 7) + 2(4x + 1)$ — $20x + 30$

45
Simplify. $(5x + 2) + 3(x + 4)$ — $8x + 14$

FEEDBACK UNIT 6

This quiz reviews the preceding unit. Answers are at the back of the book.

Simplify.

1. $6(x + 4)$
2. $3(4x + 7)$
3. $(2x + 9)$
4. $5(x + 6)$
5. $3(2x + 5)$
6. $2x + 3(5 + x)$
7. $4 + 7(2x + 3) + x$
8. $2 + (4x + 3) + 7x$
9. $5(9x + 1) + 2(x + 3)$
10. $8(x + 2) + (3x + 5)$

UNIT 7: APPLICATIONS

In this Applications Section, the format of the text has been altered. Answers for the problems appear beneath them rather than in the right-hand column. Your studying emphasis should be on learning the best procedures to follow with word problems. For that reason, once the procedure is learned a calculator may be used to complete the answer.

1

George has 1.05 kilograms of aspirin in one bottle and 0.98 more kilograms of aspirin in another. How much aspirin does he have in the two bottles?

Answer: The clue word in the problem is **more**. It often indicates that the problem can be solved by addition.

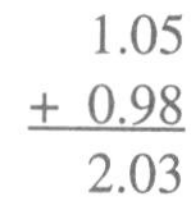

$$\begin{array}{r} 1.05 \\ +\ 0.98 \\ \hline 2.03 \end{array}$$

2.03 kilograms is the total amount of aspirin George has in the two bottles.

2
One book weighs 28.3 kilograms and another book weighs 12.1 kilograms less. What is the weight of the second book?

Answer: The word **less** is the clue to solving the problem because it indicates subtraction.

$$\begin{array}{r} 28.3 \\ -\ 12.1 \\ \hline 16.2 \end{array}$$

16.2 kilograms is the weight of the second book.

3
Find 0.8 of 53.

Answer: To find "0.8 of 53" the word "**of**" indicates that 0.8 should be multiplied by 53.

$0.8 \text{ of } 53 = 0.8 \bullet 53 = 42.4$

The answer is 42.4

4
Cola is sold by the case or by the can, but the cost is the same. If each can costs $.60 and a case costs $14.40, find the number of cans in a case.

Answer: The word **each** is a clue to solving the problem because the word usually requires multiplication or division. In this case 14.40 should be divided by .60.

$\$14.40 \div \$.60 = 24$

There are 24 cans in a case.

5
Umco Manufacturing has increased by $1.37 the unit price of one of its products. If the old price was $157.19, what is the new price?

Answer: The word **increased** is the clue to solving the problem because it indicates the problem requires addition.

$$\begin{array}{r} 157.19 \\ +\ 1.37 \\ \hline 158.56 \end{array}$$

$158.56 is the new unit price.

6

Jose has $1,268.09 in his savings account and Maria has $932.78 in hers. How much more is in Jose's account than in Maria's?

Answer: The phrase **how much more** is the clue to solving the problem because it indicates subtraction is needed.

$$\begin{array}{r} 1{,}268.09 \\ -\ 932.78 \\ \hline 335.31 \end{array}$$

Jose has $335.31 more than Maria.

7

If 25 radios cost $9.50 each, what is the cost of all the radios?

Answer: The word **each** is a clue that the problem requires multiplication.

$$25 \bullet 9.50 = 237.50$$

$237.50 is the cost of the 25 radios.

8

The Allens need to fence in their back yard so their dog cannot get loose. The yard is rectangular with length 45 feet and width 22 feet. How many feet of fencing material will be needed?

	45	
22		22
	45	

Answer: Draw a picture of the problem situation. Notice that the picture shows the rectangle has two lengths and two widths. The problem requires **finding the perimeter** of the rectangle and the distances should be added.

$$\begin{array}{r} 45 \\ 45 \\ 22 \\ +\ 22 \\ \hline 134 \end{array}$$

134 feet of fencing material is needed.

FEEDBACK UNIT 7 FOR APPLICATIONS

1. The ingredients in a medicine tablet weighed 2.03 mg, 14 mg, 8.785 mg and 0.573 mg. Find the sum of the weights of the ingredients.

2. A bank acount had $925.43 in it before $156.87 was withdrawn. How much was left?

3. Find the perimeter of a rectangle which has length 3.47 inches and width of 2.39 inches.

4. Tom has taken four math tests and received scores of 92, 85, 67, and 82. What is his average carried to the first decimal place?

5. 0.05 of a vat of liquid is sediment. If the vat holds 5,000 liters, how many liters are sediment?

SUMMARY FOR THE ALGEBRA OF THE COUNTING NUMBERS

The following mathematical terms are crucial to an understanding of this chapter.

- Numerical expression
- Variable
- Equivalent expressions
- Addends
- Associative Law of Addition
- Commutative Law of Multiplication
- Associative Law of Multiplication
- Distributive Law of Multiplication over Addition
- Open expression
- Evaluating open expressions
- Commutative Law of Addition
- Substitution
- Simplifying open expressions
- Factors
- Multiplication Law of One
- Like terms
- Removing parentheses

In Chapter 2 the algebra of the counting numbers was begun. The set of counting numbers {1, 2, 3, . . . } offers a simple example of many of the properties that are important in the study of algebra. At this point it is important that the following laws be understood.

a. Commutative Law of Addition
b. Associative Law of Addition
c. Commutative Law of Multiplication
d. Associative Law of Multiplication
e. Multiplication Law of One
f. Distributive Law of Multiplication over Addition

Variables such as x and y are used to represent places for any counting number. The expression $2 + x + 3$ can be simplified to $x + 5$, which means that the open sentence $2 + x + 3 = x + 5$ is true for any counting-number replacement of x.

Chapter 2 Mastery Test

The following questions test the objectives of Chapter 2. Answers are at the back of the book. The number in parentheses which follows each problem indicates the unit in which it can be learned.

Evaluate the following expressions.

1. $6 + 4x$ when $x = 5$ (2)

2. $8(2 + x)$ when $x = 7$ (2)

3. $3x + 5y$ when $x = 4$ and $y = 3$ (2)

4. $x(5 + x)$ when $x = 3$ (2)

Simplify the following expressions.

5. $4 + x + 11$ (3)

6. $5x \bullet 7$ (4)

7. $6y + 2(5y + 3)$ (5)

8. $7(3 + 2x)$ (5)

9. $8x + 1 + 6x + 4$ (5)

10. $4 + 3x + 5(2x + 6)$ (6)

11. $(x + 7) + 3(2x + 1)$ (6)

12. $3x + 4(5x + 6)$ (6)

13. $6(3 + 4x) + 5$ (6)

14. $6x + (4 + 13x)$ (6)

15. $9x + 8x + 3(x + 2)$ (6)

16. $12(x + 4) + 9(2x + 3)$ (6)

17. $6(4x + 9) + 8(2x + 7)$ (6)

18. $(2x + 13) + 3(7x + 8)$ (6)

19. $5(3x + 8) + 9(8x + 7)$ (6)

20. $2(x + 9) + (15x + 4)$ (6)

21. The seven linemen on the football team weigh 182 lbs, 202 lbs, 175 lbs, 210 lbs, 194 lbs, 183 lbs, and 198 lbs. Find their average weight. (7)

22. 1.2 of the cost of a refrigerator is the sale price. If the cost is \$300, what is the sale price? (7)

Chapter 3 Objectives

The following problems illustrate the objectives of this chapter. At this time you are not expected to know how to do these problems. However, if all of these problems are thoroughly understood, proceed directly to the Chapter 3 Mastery Test. The number in parentheses which follows each problem indicates the unit in which it can be learned.

1. What counting-number replacement for x will make the open sentence $x + 7 = 19$ a true statement? (1)

2. What counting-number replacement for x will make the open sentence $4x = 28$ a true statement? (1)

3. Solve the equation. $x + 17 = 23$ (1)

4. Solve the equation. $7x = 56$ (1)

5. Solve. $9 + x = 18$ (1)

6. Solve. $32 = 8x$ (1)

7. Solve. $x + 4x = 35$ (2)

8. Solve. $36 = 2(3x)$ (2)

9. Solve. $2(5x) = 2 + 7(4)$ (2)

10. Solve. $5x + 2(3x) = 5(3) + 7$ (2)

Using the set of counting numbers, find the truth set for each of the following equations.

11. $8 + x + 13 = 27$ (3)

12. $3 + x + 7 = 10$ (3)

13. $x + 12 = 16$ (3)

14. $14 + x = 8$ (3)

15. $8x = 40$ (3)

16. $32 = 6x$ (3)

17. $4(2x + 3) = 12 + 8x$ (3)

18. Solve and check. $12 + x + 7 = 31$ (4)

19. Solve and check. $5x + 3(3x) = 42$ (4)

20. Solve and check. $4(2x) + x = 45$ (4)

Chapter 3
Solving Equations with the Counting Numbers

Unit 1: Statements and Open Sentences

The following mathematical terms are crucial to an understanding of this unit.

Mathematical statement	Open sentence
Equation	Solving an equation

1
In words, $7 + 3 = 2 \cdot 5$ may be translated as:
The sum of 7 and 3 is equal to the product of 2 and 5.
Is the claim of $7 + 3 = 2 \cdot 5$ true or is it false? True

2
In words, $4 + 9 = 3 \cdot 5$ may be translated as:
The sum of 4 and 9 is equal to the product of 3 and 5.
Is the claim of $4 + 9 = 3 \cdot 5$ true or is it false?

False

3
Both $7 + 3 = 2 \cdot 5$ and $4 + 9 = 3 \cdot 5$ are **mathematical statements** because either can be judged true or false. Is $5 + 3[2 \cdot 4 + 7]$ a mathematical statement?

No. It makes no claim that can be judged true or false.

4
Is the following a mathematical statement? $4 + 7 = 11$

Yes, a true one

5
Is the following a mathematical statement? $4 \cdot 5 = 9$

Yes, a false one

6
Neither $2 + x = 7$ or $5x = 20$ are mathematical statements because neither can be judged true or false until x is replaced by a number. $2 + x = 7$ is an **open sentence** and will become a mathematical statement only when x is replaced by a number. Is $5x = 20$ also an open sentence?

Yes

7
Open sentences are neither true nor false. If x is replaced by 7 in $8 + x = 13$, a mathematical statement is obtained. Is the statement $8 + 7 = 13$ true?

No, it is false

8
If x is replaced by 2 in the open sentence $5x + 3 = 13$, a mathematical statement is obtained. Is $5 \cdot 2 + 3 = 13$ true?

Yes, it is true

9
If x is replaced by 2 in $6x + 2 = 19$, is the result true?

No, $6 \cdot 2 + 2 = 19$ is false

10
For the open sentence $5x = 20$, find a replacement for x that will make a true statement. 4

11
For the open sentence $x + 12 = 18$, find a replacement for x that will make a true statement. 6

12
For the open sentence $3y = 36$, find a replacement for y that will make a true statement. 12

Open sentences which use the equal sign (=) are called **equations**. An equation is solved by finding the replacement(s) for the variable that will make the equation a true statement.

13
An open sentence such as $5x + 3 = 19$ is often referred to as an **equation**. $6x + 2 = 30$ is an open sentence or an ______. equation

14
Find the counting-number replacement for y in the equation $y + 7 = 9$ that will make the statement true. 2

15
What counting-number replacement for r in the equation $3 + r = 10$ will make a true statement? 7

16
To **solve the equation** $x + 3 = 7$ is to find the counting-number replacement for x that will make a true statement. Solve the equation $x + 3 = 7$. 4

17
Solve the equation. $5 + x = 13$ 8

18
Solve the equation. $x + 3 = 19$ 16

19
Solve. $13 + x = 17$ 4

20
Solve. $9 = x + 4$ 5

21
Solve. $47 = x + 9$ 38

22
Solve. $23 = x + 14$ 9

23
Solve. $45 + x = 50$ 5

24
Solve. $25 + x = 35$ 10

25
Solve. $18 + x = 24$ 6

26
Solve. $17 + x = 20$ 3

27
Solve. $2x = 10$ 5

28
Solve. $3x = 33$ 11

29
Solve. $36 = 9x$ 4

30
Solve. $6x = 42$ 7

31
Solve. $10x = 150$ 15

32
Solve. $63 = 7x$ 9

33
Solve. $8x = 64$ 8

34
Solve. $x + 15 = 23$ — 8

35
Solve. $5x = 40$ — 8

36
Solve. $5x = 80$ — 16

37
Solve. $4 + x = 17$ — 13

38
Solve. $3x = 9$ — 3

39
Solve. $7x = 42$ — 6

40
Solve. $24 = 4x$ — 6

41
Solve. $8x = 24$ — 3

42
Solve. $32 = 4x$ — 8

43
Solve. $13 + x = 18$ — 5

44
Solve. $12x = 180$ — 15

FEEDBACK UNIT 1

This quiz reviews the preceding unit. Answers are at the back of the book.

1. Which of the following is/are mathematical statements?
   a. $9 + 3 \cdot 5 = 24$
   b. $3x = 18$
   c. $4 \cdot 7 + 8 = 5$
   d. $6 + 1 = 7$
2. Which of the following is/are open sentences?
   a. $5 + 3 \cdot 2 = 17$
   b. $4x = 19$
   c. $x + 41 = 65$
   d. $8 + 13 = 3x$

Solve each of the following.

3. $x + 13 = 91$
4. $7x = 56$
5. $19 = x + 5$
6. $48 = 19 + x$
7. $63 = 21x$
8. $17 + x = 31$
9. $17x = 51$
10. $x + 12 = 455$

UNIT 2: SOLVING EQUATIONS

The following mathematical term is crucial to an understanding of this unit.

Solution

1

To solve $x + 7 + 3 = 19$, first notice that the expression on the **left side of the equal sign** (=) can be simplified.

$x + 7 + 3 = 19$ — $x + 7 + 3$ simplifies
$x + 10 = 19$ — to $x + 10$

The **solution** for $x + 10 = 19$ is 9.

Solve $2 + r + 7 = 12$ by first simplifying $2 + r + 7$.

$r + 9 = 12$
$r = 3$

2
Solve. $8 + z + 5 = 21$

$z + 13 = 21$
Solution is 8

3
Solve. $3x + 2x = 30$

$5x = 30$
Solution is 6

4
Solve. $27 = 7x + 2x$

$27 = 9x$
Solution is 3

5
Solve. $36 = 8x + 4x$

3

6
Solve. $x + 2x = 12$

4

7
Solve $2(3x) = 4 \cdot 7 + 2$ by first simplifying both sides of the equation.

$6x = 30$
Solution is 5

8
Solve. $(5x)2 = 8 \cdot 7 + 4$

$10x = 60$
Solution is 6

9
Solve. $7x + x = 24$

3

10
Solve. $4(3x) = 24$

2

11
Solve. $5(3x) = 45$

3

12
Solve. $2(4x) = 40$

5

13
Solve. $3x + 2(5x) = 13$

1

14
Solve. $x + x = 8$

4

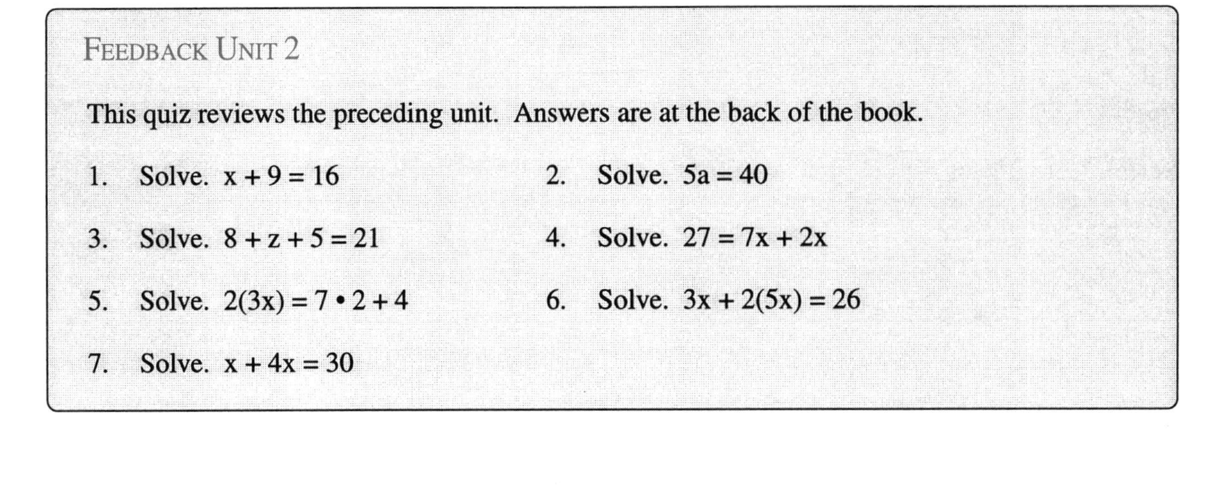

FEEDBACK UNIT 2

This quiz reviews the preceding unit. Answers are at the back of the book.

1. Solve. $x + 9 = 16$
2. Solve. $5a = 40$
3. Solve. $8 + z + 5 = 21$
4. Solve. $27 = 7x + 2x$
5. Solve. $2(3x) = 7 \cdot 2 + 4$
6. Solve. $3x + 2(5x) = 26$
7. Solve. $x + 4x = 30$

UNIT 3: FINDING TRUTH SETS FOR EQUATIONS USING THE SET OF COUNTING NUMBERS

The following mathematical terms are crucial to an understanding of this unit.

Truth set Replacement set Empty set

1

Using the set of counting numbers, {5} is the **truth set** of $4x = 20$ because 5 is the only counting number that will make $4x = 20$ a true statement. Find the truth set of $7x = 21$.

{3}

2

Using the set of counting numbers as the **replacement set**, determine the truth set of $x + 3 = 22$.

{19}

3

Using the set of counting numbers {1, 2, 3, . . . }, find the truth set for $6x + 3x = 36$.

{4}

4
Using the set {1, 2, 3, . . . }, find the truth set for $x + 4x = 500$.

{100}

5
Using the set of counting numbers {1, 2, 3, . . . }, find the truth set for $200 + x = 283$.

{83}

6
Using the set {1, 2, 3, . . . }, find the truth set for $493 = x + 93$.

{400}

7
Using the set of counting numbers, find the truth set for $5(2x) = 170$.

{17}

8
Using the set of counting numbers, find the truth set for $2x = 480$.

{240}

9
Some equations have more than one element in their truth sets. The equation $1x = x$ becomes a true statement for any counting-number replacement of x. Will the equation $x + 4 = 4 + x$ become a true statement for any counting-number replacement of x?

Yes

10
The truth set of $1x = x$ is {1, 2, 3, . . . } because every counting-number replacement makes the statement true. What is the truth set of $x + 4 = 4 + x$?

{1, 2, 3, . . . }

11
The truth set of $9(x + 4) = 9x + 36$ is {1, 2, 3, . . . } because $9(x + 4)$ and $9x + 36$ are equivalent expressions. What is the truth set of $5(x + 4) = 5x + 20$?

{1, 2, 3, . . . }

12
Since $5x$ and $2x + 3x$ are equivalent expressions, the truth set of $5x = 2x + 3x$ is {1, 2, 3, . . . }. What is the truth set of $9x = 7x + 2x$?

{1, 2, 3, . . . }

13
Using {1, 2, 3, . . . }, find the truth set for $2x = 24$.

{12}

14
Using {1, 2, 3, . . . }, find the truth set for $5 + x = 23$.

{18}

15
Using {1, 2, 3, . . . }, find the truth set for $7x + x = 8x$.

{1, 2, 3, . . . }

16
Using {1, 2, 3, . . . }, find the truth set for $3x = 75$.

{25}

17
Using {1, 2, 3, . . . }, find the truth set for $x + 8 = 40$.

{32}

18
Using {1, 2, 3, . . . }, find the truth set for $6(x + 7) = 42 + 6x$.

{1, 2, 3, . . . }

19
The equation $x + 5 = 3$ has no counting-number replacement that will produce a true statement because 5 is greater than 3. Is there a counting number that will make $x + 9 = 2$ a true statement?

No

20
The equation $7x = 10$ has no counting number that will produce a true statement because 10 cannot be obtained by multiplying 7 by any counting number. Is there a counting number that will make $5x = 13$ a true statement?

No

21
The truth set of $x + 5 = 3$ is the **empty set**, { }, because no counting-number replacement will make the equation a true statement. What is the truth set of $5x = 13$?

{ }

22
The truth set of $x + 7 = 13$ is {6}. The truth set of $5 + x + 3 = x + 8$ is {1, 2, 3, . . . }. What is the truth set of $x + 12 = 5$?

{ }

23
The truth set of $3x = 15$ is $\{5\}$. The truth set of $3(x + 4) = 3x + 12$ is $\{1, 2, 3, \ldots\}$. What is the truth set of $2x = 11$?

$\{\ \}$

24
Find the truth set for $14 + x = 9$.

$\{\ \}$

25
Find the truth set for $4x = 13$.

$\{\ \}$

26
Using $\{1, 2, 3, \ldots\}$, find the truth set for $x + 9 = 14$.

$\{5\}$

27
Using the set of counting numbers, find the truth set for $3x = 8$.

$\{\ \}$

28
Using the set of counting numbers, find the truth set for $3 + x + 14 = x + 17$.

$\{1, 2, 3, \ldots\}$

29
Using the set of counting numbers, find the truth set for $x + 19 = 2$.

$\{\ \}$

30
Using the set of counting numbers, find the truth set for $7x = 21$.

$\{3\}$

31
Using $\{1, 2, 3, \ldots\}$, find the truth set for $4x = 23$.

$\{\ \}$

32
Using the set of counting numbers, find the truth set for $x + 4 = 3$.

$\{\ \}$

33
Using the set of counting numbers, find the truth set for $12 + x = 21$.

$\{9\}$

34
Using the set of counting numbers, find the truth set for $x + 8 = 3$.

$\{\ \}$

35
Using the set of counting numbers, find the truth set for $9x = 72$.

$\{8\}$

36
Using the set of counting numbers, find the truth set for $7(x + 5) = 7x + 35$.

$\{1, 2, 3, \ldots\}$

37
Using the set of counting numbers, find the truth set for $7x = 18$.

$\{\ \}$

38
Using the set of counting numbers, find the truth set for $x + 12 = 21$.

$\{9\}$

39
Using the set of counting numbers, find the truth set for $17 = 4 + x$.

$\{13\}$

40
Using the set of counting numbers, find the truth set for $9x = 63$.

$\{7\}$

41
Using the set of counting numbers, find the truth set for $48 = 6x$.

$\{8\}$

42
Using the set of counting numbers, find the truth set for $6x + 5x = 44$.

$\{4\}$

43
Using the set of counting numbers, find the truth set for $3 + 9 + x = 12 + x$.

$\{1, 2, 3, \ldots\}$

44
Using the set of counting numbers, find the truth set for $19 + x = 10$.

$\{\ \}$

45
Using the set of counting numbers, find the truth set for $2x + 10 = 2(x + 5)$.

$\{1, 2, 3, \ldots\}$

46
Using the set of counting numbers, find the truth set for $36x = 72$.

{2}

47
Using the set of counting numbers, find the truth set for $11x = 32$.

{ }

48
Using the set of counting numbers, find the truth set for $x + 13 = 14$.

{1}

49
Using the set of counting numbers, find the truth set for $9 + x = 4$.

{ }

50
Using the set of counting numbers, find the truth set for $x + 17 = 23$.

{6}

51
Using the set of counting numbers, find the truth set for $15x = 60$.

{4}

52
Using the set of counting numbers, find the truth set for $31 = 6x$.

{ }

53
Using the set of counting numbers, find the truth set for $x + 54 = 12$.

{ }

54
Using the set of counting numbers, find the truth set for $7x = 77$.

{11}

55
Using the set of counting numbers, find the truth set for $12 + x = 27$.

{15}

56
Using the set of counting numbers, find the truth set for $3(2x + 5) = 15 + 6x$.

{1, 2, 3, . . . }

FEEDBACK UNIT 3

This quiz reviews the preceding unit. Answers are at the back of the book.

Using the set of counting numbers, find truth sets for the following equations.

1. $x + 8 = 14$
2. $2x = 22$
3. $6 + x + 3 = x + 9$
4. $x + 5 = 19$
5. $x + 8 + 5 = 13 + x$
6. $3x = 27$
7. $x + 6 = 1$
8. $3x = 10$

UNIT 4: SOLVING EQUATIONS AND CHECKING THE SOLUTIONS

The following mathematical term is crucial to an understanding of this unit.

Checking a solution

1

The equation $4x = 12$ has 3 as its solution. To check 3 as the solution, 3 is substituted for x in the original equation. If the substitution of 3 for x makes a true statement, the solution has checked.

When $x = 3$, then $4x = 12$ becomes $4 \cdot 3 = 12$. Since $4 \cdot 3 = 12$ or $12 = 12$ is true, the solution checks.

Solve and **check the solution**. $7x = 21$

$7x = 21$
$x = 3$
Check:
$7 \cdot 3 = 21$
$21 = 21$ is true.

2

The equation $x + 8 = 13$ has 5 as its solution. Check the solution by substituting 5 for x in $x + 8 = 13$. Show that the substitution will make a true statement.

When $x = 5$, then $x + 8 = 13$ becomes $5 + 8 = 13$.

Since $5 + 8 = 13$ or $13 = 13$ is true, the solution checks.

Solve and check the equation. $x + 3 = 11$

$x + 3 = 11$
$x = 8$
Check:
$8 + 3 = 11$
$11 = 11$ is true.

3

To solve the equation $7 + x + 8 = 19$, first simplify the left side of the equation.

$$7 + x + 8 = 19$$
$$x + 15 = 19$$
$$x = 4$$

To check the solution, 4 is substituted in the original open sentence.

$$7 + x + 8 = 19$$
$$7 + 4 + 8 = 19$$
$$11 + 8 = 19$$
$$19 = 19 \text{ is true}$$

Solve and check. $3 + x + 6 = 14$

$x = 5$
Check:
$3 + 5 + 6 = 14$
$14 = 14$ is true.

4

To solve the equation $3x + 3(4x) = 30$, first simplify the left side of the equation.

$$3x + 3(4x) = 30$$
$$3x + 12x = 30$$
$$15x = 30$$
$$x = 2$$

To check the solution, 2 is substituted in the original equation.

$$3x + 3(4x) = 30$$
$$3 \cdot 2 + 3(4 \cdot 2) = 30$$
$$6 + 3 \cdot 8 = 30$$
$$6 + 24 = 30$$
$$30 = 30 \text{ is true}$$

Solve and check. $5x + 2(3x) = 44$

$x = 4$
Check:
$5 \cdot 4 + 2(3 \cdot 4) = 44$
$20 + 2(12) = 44$
$20 + 24 = 44$
$44 = 44$ is true.

5
Solve and check. $3x + 2x = 35$

$x = 7$
Check:
$3 \cdot 7 + 2 \cdot 7 = 35$
$21 + 14 = 35$
$35 = 35$ is true.

6
Solve and check. $5 + x = 17$

$x = 12$
Check:
$5 + 12 = 17$
$17 = 17$ is true.

7
Solve and check. $3(2x) + x = 42$

$x = 6$
Check:
$3(2 \cdot 6) + 6 = 42$
$3(12) + 6 = 42$
$36 + 6 = 42$
$42 = 42$ is true.

8
Solve and check. $x + 2x = 18$

$x = 6$
Check:
$6 + 2(6) = 18$
$6 + 12 = 18$
$18 = 18$ is true.

9
Solve and check. $x + 13 = 31$

$x = 18$
Check:
$18 + 13 = 31$
$31 = 31$ is true.

10
Solve and check. $3(2x) + 5(2x) = 48$

$x = 3$
Check:
$3(2 \cdot 3) + 5(2 \cdot 3) = 48$
$3(6) + 5(6) = 48$
$18 + 30 = 48$
$48 = 48$ is true.

11

Suppose that $5x = 40$ were solved and the solution was found to be 7. The check would show that the solution was unacceptable in the following way.
To check the solution, 7 is substituted in the original equation.

$$5x = 40$$
$$5 \cdot 7 = 40$$
$$35 = 40 \text{ is } \textbf{false}$$

Check $x = 5$ as a solution for $4(3x) = 8 \cdot 6 + 4$.

Check:
$4(3 \cdot 5) = 8 \cdot 6 + 4$
$4(15) = 48 + 4$
$60 = 52$ is false.

FEEDBACK UNIT 4

This quiz reviews the preceding unit. Answers are at the back of the book.

Solve and check the following equations.

1. $3 + x = 14$
2. $7x = 56$
3. $x + 5 = 2 + 3(4)$
4. $6(2x) + x = 3(10) + 9$
5. $x + x = 5(3) + 3$

UNIT 5: APPLICATIONS

In this Applications Section, the format of the text has been altered. Answers for the problems appear beneath them rather than in the right-hand column. Your studying emphasis should be on learning the best procedures to follow with word problems. For that reason, once the procedure is learned a calculator may be used to complete the answer.

1

75% of a test with 60 questions are true-false questions. How many are true-false questions?

Answer: Percent problems always involve three numbers: part (p), base (b), and rate (r). The numbers are related by the formulas p = rb, r = p ÷ b, and b = p ÷ r. These three formulas are shown in the figure below. To solve this problem, note that it seeks **part** of the problems on the test. Hence, p = rb.

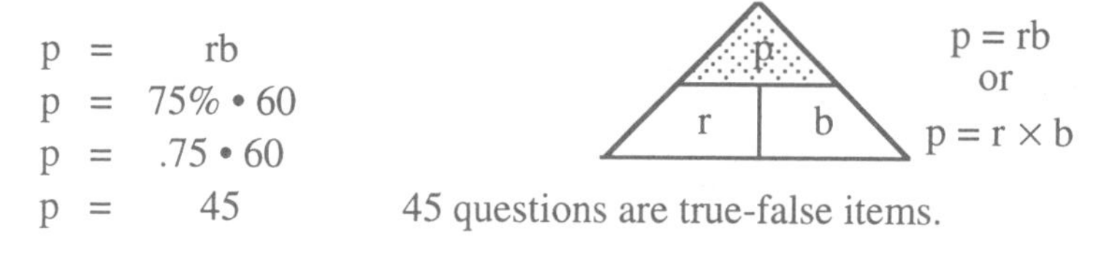

45 questions are true-false items.

2

22% of a solution of rubbing alcohol is water. There are 3.96 liters of water in the solution. How many liters are in the solution?

Answer: Since 3.96 liters is part of the solution, this problem requires finding the **base (b)**. The figure below shows a way to remember the formula.

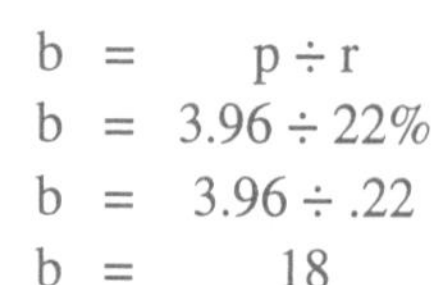

p, r, b

$b = \frac{p}{r}$
or
b = p ÷ r

There are 18 liters in the solution.

3
A class of 45 students has 19 boys.
What percent of the class is boys?

Answer: To "find the percent," means to find the rate (r). The figure below will help remember the relationship.

$r = 19 \div 45$
$r = 42\frac{2}{9}$

To the nearest one percent r is 42%.

4
Last year Mr. Gottza paid $27,000 for his income taxes which was 38% of his income. Which of the following questions fits the situation of Mr. Gottza?
38% of _____ is $27,000.
38% of $27,000 is _____.

Answer: The three numbers in a percent problem need to be properly designated. To help in this process, learn to restate the problem in the form: **r** of **b** is **p**

In Gottza's situation: 38% of income is taxes, or
38% of _______ is $27,000.

5
Does the question below ask for base, rate, or part?
What percent of a ring weighing 20 grams is silver if there are 8 grams of silver in the ring?

Answer: The rate (r) is the most easily recognized number in a percent problem because it is accompanied by the % symbol or the word "percent."

Since this problem asks for the percent, it is the rate that is sought.

6

What percent of a ring weighing 20 grams is silver if there are 8 grams of silver in the ring?
Which of the following statements fits the situation above?
r% of 20 is 8.
r% of 8 is 20.

Answer: The **base** of a percent problem is usually preceded by the word "of" and if the problem is restated, will always be in that position.

The problem has the phrase: What percent **of** the ring
Find the number that describes the weight of the ring and that is the base.
The base is 20 and the correct restatement is: r% of 20 is 8.

7

2% of the students in a school have a hearing impairment. There are 1500 students in the school. Which of the following is an accurate restatement of the situation.
2% of 1500 is _____.
2% of _____ is 1500.

Answer: The problem has the phrase: 2% **of** the students
Since 1500 describes the students at the school, the correct choice is:
2% of 1500 is _____.

8

How many girls are in a class of 450 students if 52% of the class is girls? Does the question ask for the base, rate, or part?

Answer: Two possible restatements of the situation can be investigated.

52% of the girls is 450. or 52% of 450 is the girls.

The second choice is correct. This means the base is 450 and the original question asks for the part.

9

A store makes a 6% profit on its total sales of $2,000,000. Identify 6% and $2,000,000 as the base, rate, or part.

Answer: 6% is the rate because it has the percent symbol, %. The phrase "total sales of $2,000,000" indicates the base is $2,000,000.

FEEDBACK UNIT 5 FOR APPLICATIONS

1. Last year Mr. Gottza paid $27,000 for his income taxes which was 38% of his income. What was his income?

2. What percent of a ring weighing 20 grams is silver if there are 8 grams of silver in the ring?

3. 2% of the students in a school have a hearing impairment. There are 1500 students in the school. Find the number of students with a hearing impairment.

4. How many girls are in a class of 450 students if 52% of the class is girls?

5. A store makes a 6% profit on its total sales of $2,000,000. Find the profit.

SUMMARY FOR SOLVING EQUATIONS WITH COUNTING NUMBERS

The following mathematical terms are crucial to an understanding of this chapter.

Mathematical statement	Open sentence
Equation	Solving an equation
Left side of the equals sign	Solution of an equation
Truth set	Replacement set
Empty set	Checking a solution

To solve or find the truth set of an equation such as $x + 5 = 7$ means to find the set of all counting-number replacements for x that will make the statement true.

Some equations, such as $x + 5 = 2$ and $3x = 10$, do not have a counting-number replacement that will produce a true statement. The solution for these equations using the set of counting numbers is, therefore, the empty set, { }.

The truth set for $2 + x + 3 = x + 5$ is {1, 2, 3, . . . }, because every counting number makes the open sentence $2 + x + 3 = x + 5$ a true statement.

Many equations have exactly one counting number that will make them true statements. For example, $x + 17 = 20$ is an equation that has exactly one counting-number solution, and that is 3.

Some equations do not have a counting number solution. For example, there is no counting number replacement for x that will make $x + 17 = 11$ a true statement. Using the set of counting numbers, the equation has { } as its truth set.

To check an equation, its proposed solution must be substituted for the letter in the original open sentence. If the resulting mathematical statement is true, the check is completed.

CHAPTER 3 MASTERY TEST

The following questions test the objectives of Chapter 3. Answers are at the back of the book. The number in parentheses which follows each problem indicates the unit in which it can be learned.

1. What counting-number replacement for x will make the open sentence $x + 23 = 41$ a true statement? (1)
2. What counting-number replacement for x will make the open sentence $9x = 72$ a true statement? (1)
3. Solve. $x + 14 = 29$ (1)
4. Solve. $6x = 54$ (1)
5. Solve. $17 + x = 28$ (1)
6. Solve. $36 = 9x$ (1)
7. Solve. $6x + 8x = 84$ (2)
8. Solve. $75 = 3(5x)$ (2)
9. Solve. $3(8x) = 2 + 10(7)$ (2)
10. Solve. $2x + 3(4x) = 6(8) + 22$ (2)
11. Solve and check. $14 + x + 31 = 51$ (2)
12. Solve and check. $3x + 6(2x) = 7(8) + 4$ (2)
13. Solve and check. $4 + 3(5) = 5(2x) + 9x$ (2)

For problems 14-20, use the set of counting numbers and find the truth set for each of the equations.

14. $9 + x + 8 = 24$ (3)
15. $x + x = 3 + 5(5)$ (3)
16. $4 + x + 11 = 15 + x$ (3)
17. $3x = 13$ (3)
18. $5 + x + 7 = 9$ (3)
19. $4(3x + 2) = 8 + 12x$ (3)
20. $8x + 3(5x) = 9 + 5(12)$ (3)
21. Solve and check. $x + 11 = 13$ (4)
22. Solve and check. $7x = 42$ (4)
23. 87% of milk is water. If a bottle of milk contains 2.61 liters of water, how much milk is in the bottle? (5)
24. A bottle of medicine contains 12 liters. 2.6% of the contents is active ingredients. How much of the medicine is active ingredients? (5)

Chapter 4 Objectives

The following problems illustrate the objectives of this chapter. At this time you are not expected to know how to do these problems. However, if all these problems are thoroughly understood, proceed directly to the Chapter 4 Mastery Test. The number in parentheses which follows each problem indicates the unit in which it can be learned.

1. What is the opposite of -13? (1)
2. Solve. $-5 + x = 0$ (1)
3. Evaluate. $-12 + (-7) =$ ______ (2)
4. Evaluate. $6 - (-3) =$ ______ (2)
5. Evaluate. $-3 - 9 =$ ______ (2)
6. Evaluate. $-8 \cdot 9 =$ ______ (3)
7. Evaluate. $-9 \cdot (-7) =$ ______ (3)
8. Evaluate. $6 \cdot -7 =$ ______ (3)
9. Evaluate. $-4 \cdot (-8) =$ ______ (3)
10. Evaluate. $-7 - 4 - 8 =$ ______ (4)
11. Evaluate. $3 + (-5) - 4 =$ ______ (4)
12. Evaluate. $3 - (-5) + (-3) =$ ______ (4)
13. Evaluate. $8 - 3 - 5 =$ ______ (4)
14. Evaluate. $-4(7 \cdot -3) =$ ______ (5)
15. Evaluate. $6(-3) \cdot -2 =$ ______ (5)
16. Evaluate. $(-2)^4 =$ ______ (5)
17. Evaluate. $-2^4 =$ ______ (5)
18. Evaluate. $3^4 =$ ______ (5)
19. List the first five composite numbers. (6)
20. List the prime numbers between 16 and 23. (6)

CHAPTER 4

THE ARITHMETIC OF THE INTEGERS

UNIT 1: THE SET OF INTEGERS

The following mathematical terms are crucial to an understanding of this unit.

Opposite	Negative
Negative integers	Integers
Set of integers	Positive integers

The equation $x + 4 = 1$ cannot be made into a true statement by any counting number, because when 4 is added to a counting number the sum is always greater than 1.

In this unit a new set of numbers will be studied to provide numbers that will make equations such as $x + 4 = 1$ true statements. This new set of numbers is called the set of integers.

"Integer" is pronounced in′-te-jer.

The set of integers consists of all the counting numbers, zero, and the opposite of each counting number.

1
There is no counting-number replacement for x in the open sentence $x + 2 = 2$ that will make the statement true. (Zero is not a counting number.) If {0, 1, 2, . . . } is used, the truth set of $x + 2 = 2$ is {0}. Use the set {0, 1, 2, . . . } to find the truth set of $x + 5 = 5$.

{0}

2
Using {0, 1, 2, . . . }, find the truth set of $x + 19 = 19$.

{0}

3
Using the set {0, 1, 2, . . . }, find the truth set of $4 + x = 4$.

{0}

4
Using the set {0, 1, 2, . . . }, find the truth set of $x + 3 = 5$.

{2}

5
Using {0, 1, 2, . . . }, find the truth set for $x + 73 = 73$.

{0}

6
Equations such as $x + 1 = 0$, $x + 2 = 0$, and $x + 3 = 0$ have no solution in the set {0, 1, 2, . . . }. Does $x + 4 = 0$ have a solution in the set {0, 1, 2, . . . }?

No

7
To have a replacement for x in the open sentence $x + 1 = 0$ it is necessary to have a new number, which is called the **opposite** of 1 or **negative** 1. To have a replacement for x in the open sentence $x + 2 = 0$ it is necessary to have a new number, which is called the ______ of 2 or ______ 2.

opposite
negative

8
The opposite of 1 is shown as -1. The opposite of 2 is shown as -2. Write the opposite of 3.

-3

9
-17 is the opposite of 17. Write the opposite of 23.

-23

10
71 is the opposite of -71. Write the opposite of -43.

43

11
The opposite of -27 is ______. 27

12
9 is the opposite of ______. -9

13
27 is the opposite of ______. -27

14
-15 is the opposite of 15 and -162 is the opposite of 162. Does every counting number have an opposite? Yes

15
The set of all opposites of the counting numbers would be shown by the set {-1, -2, -3, . . . }. Is -12 in the set {-1, -2, -3, . . . }? Yes

16
Is -29 in the set {-1, -2, -3, . . . }? Yes

17
The set of opposites {-1, -2, -3, . . . } is called the set of **negative integers**. Every counting number has an opposite in the set of ______ integers. negative

18
-37 is in the set of ______ integers. negative

19
-942 is in the set of ______ integers. negative

The **set of integers** is { . . . -3, -2, -1, 0, 1, 2, 3, . . . }. The set contains all the **positive integers**, 1, 2, 3, etc., all the negative integers, -1, -2, -3, etc., and zero, 0, which is neither positive nor negative.

20
The set { . . . -3, -2, -1, 0, 1, 2, 3, . . . } combines the counting numbers, zero, and their opposites, the negative integers. The set is called the set of integers. Is 4 in the set of integers? Yes

21
Is 19 in the set of integers
{ . . . -3, -2, -1, 0, 1, 2, 3, . . . }? Yes

22
Is -12 in the set of integers
{ . . . -3, -2, -1, 0, 1, 2, 3, . . . }? Yes

23
Is zero in the set of integers
{ . . . -3, -2, -1, 0, 1, 2, 3, . . . }? Yes

24
Is -93 in the set of integers? Yes

25
Is 958 in the set of integers? Yes

26
Is every counting number in the set of
integers { . . . -3, -2, -1, 0, 1, 2, 3, . . . }? Yes

27
Is -37 in the set of integers? Yes

28
Is every negative integer in the set of
integers { . . . -3, -2, -1, 0, 1, 2, 3, . . . }? Yes

29
The sum of any counting number and its opposite
is zero.
1 + (-1) = 0   2 + (-2) = 0   3 + (-3) = ______ 0

30
5 + (-5) = 0 and 12 + (-12) = ______ 0

31
27 + (-27) = ______ 0

32
When a counting number and its
opposite are added, the sum is zero.
8 + (-8) = 0 and 15 + ______ = 0 -15

33
21 + (-21) = 0 and 37 + ______ = 0 -37

34

61 + ______ = 0

-61

35

The opposite of zero is itself, since 0 + 0 = 0. Zero is the only integer in the set of integers { . . . -3, -2, -1, 0, 1, 2, 3, . . . } that is its own opposite. 0 + ______ = 0.

0

36

-17 + 17 = 0 and ______ + 39 = 0

-39

37

______ and 21 = 0

-21

38

13 + (-13) = 0 and ______ + (-29) = 0

29

39

______ + (-14) = 0

14

40

56 + ______ = 0

-56

41

______ + 0 = 0

0

42

______ + (-47) = 0

47

43

84 + ______ = 0

-84

44

The truth set of x + 5 = 0 using the set of integers { . . . -3, -2, -1, 0, 1, 2, 3, . . . } is {-5}. Use the set of integers to find the truth set for x + 9 = 0.

{-9}

45

Using the set of integers, find the truth set for x + 8 = 0.

{-8}

46

Using the set of integers { . . . -3, -2, -1, 0, 1, 2, 3, . . . }, find the truth set for x + (-24) = 0.

{24}

47

Using the set of integers, find the truth set for x + 49 = 0.

{-49}

48
Using the set { . . . -3, -2, -1, 0, 1, 2, 3, . . . },
find the truth set for 21 + x = 0. {-21}

49
Using the set of integers, find the truth set
for x + (-55) = 0. {55}

50
Using the set of integers, find the truth set
for x + (-15) = 0. {15}

51
Using the set of integers, find the truth set
for 72 + x = 0. {-72}

52
Find the truth set for x + 0 = 0. {0}

53
Find the truth set for x + 418 = 0. {-418}

54
Using the set { . . . -3, -2, -1, 0, 1, 2, 3, . . . },
find the truth set for -23 + x = 0. {23}

55
In the set { . . . -3, -2, -1, 0, 1, 2, 3, . . . } the
opposites of the counting numbers are the
negative integers. The counting numbers are
the positive integers. The counting number 12
is a ______ integer. positive

56
-13 is a negative integer. 48 is a ______ integer. positive

57
34 is a ______ integer. positive

58
-33 is a ______ integer. negative

59
128 is a ______ integer. positive

60
-156 is a ______ integer. negative

61
Zero is neither a positive integer nor a negative integer. The only integer in the set { . . . -3, -2, -1, 0, 1, 2, 3, . . . } that is neither positive nor negative is ______.

0 (zero)

FEEDBACK UNIT 1

This quiz reviews the preceding unit. Answers are at the back of the book.

1. { . . . -3, -2, -1, 0, 1, 2, 3, . . . } is called the set of ______.
2. -5 is the opposite of ______.
3. 13 is the opposite of ______.
4. 0 is the opposite of ______.
5. -7 + 7 = ______
6. 47 + (-47) = ______
7. Whenever two opposites are added, their sum is ______.
8. In the set of integers the opposites of the positive integers are called the ______ integers.
9. Find the truth set for x + (-8) = 0.
10. Find the truth set for -6 + x = 0.

UNIT 2: THE NUMBER LINE FOR THE SET OF INTEGERS

The following mathematical terms are crucial to an understanding of this unit.

Number line   Addition of integers

The set of integers can easily be shown in a picture by the use of the **number line**.

-8   -6   -4   -2   0   2   4   6   8   10   12

The number line divides the set of integers into positive integers to the right of zero and negative integers to the left of zero. Zero is neither positive nor negative.

1

The set of integers is shown graphically below. The letter A marks the position of +5. The letter B marks the position of ______.

+3

2

On the number line shown below the letter C marks the position of -6 and the letter D marks the position of ______.

-2

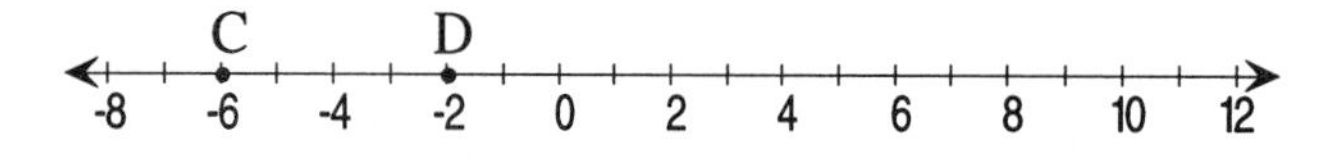

3

The positive integers lie to the right of zero on the number line below. The negative integers lie to the ______ of zero.

left

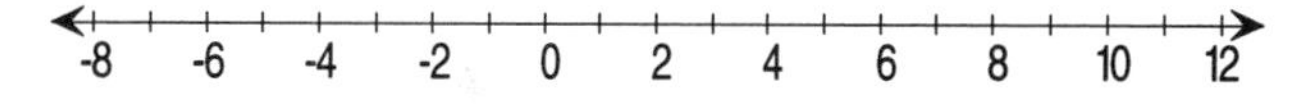

4

Two arrows are shown with the number line below. The arrow marked +6 goes to the right six units. The arrow marked -4 goes to the ______ four units.

left

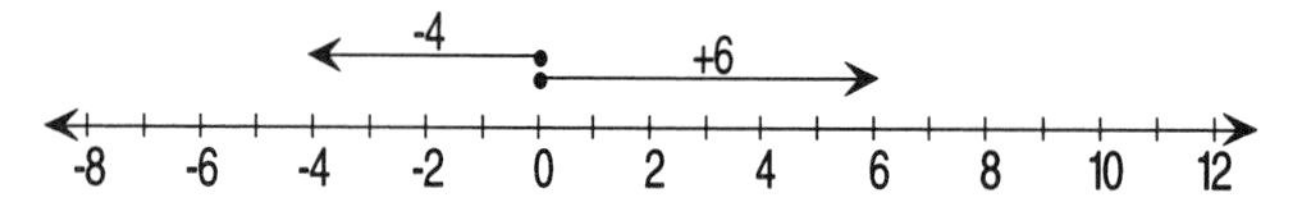

5

Positive integers represent arrows that go to the right of zero. Negative integers represent arrows that go to the ______ of zero.

left

6
The combination of +7 and +8 is shown on the number line below. +7 and +8 makes +15.

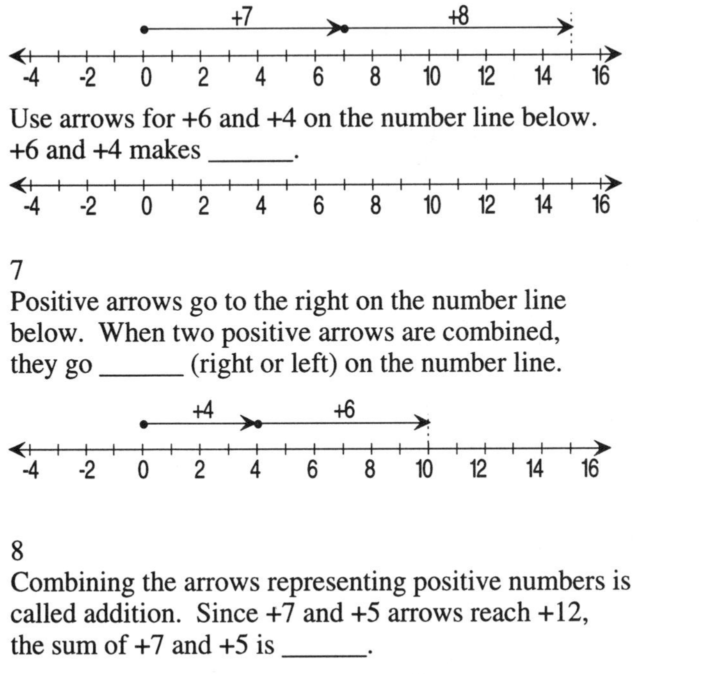

Use arrows for +6 and +4 on the number line below.
+6 and +4 makes ______.

+10

7
Positive arrows go to the right on the number line below. When two positive arrows are combined, they go ______ (right or left) on the number line.

right

8
Combining the arrows representing positive numbers is called addition. Since +7 and +5 arrows reach +12, the sum of +7 and +5 is ______.

+12

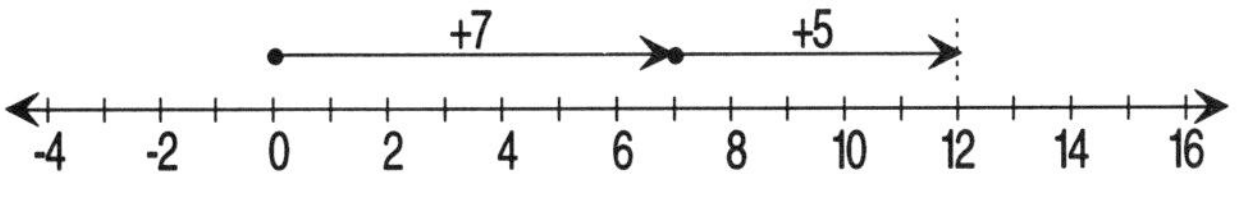

9
Combining two arrows that represent positive integers is **addition**. Since the combination of two positive integers goes to the right of the number line, the sum of two positive integers is always a ______ (positive or negative) integer.

positive

10
The sum of two counting numbers is always a counting number. The sum of two positive integers is always a ______ integer.

positive

11

$+7 + (+8) = +15$ and $+9 + (+14) =$ ______

+23

12
The sum of two positive integers is a ______ integer.

positive

The sum of two positive integers is always a positive integer.

$5 + 7 = 12$ $8 + 6 = 14$

13
The combination of -3 and -5 is shown on the number line below. -3 and -5 makes -8.

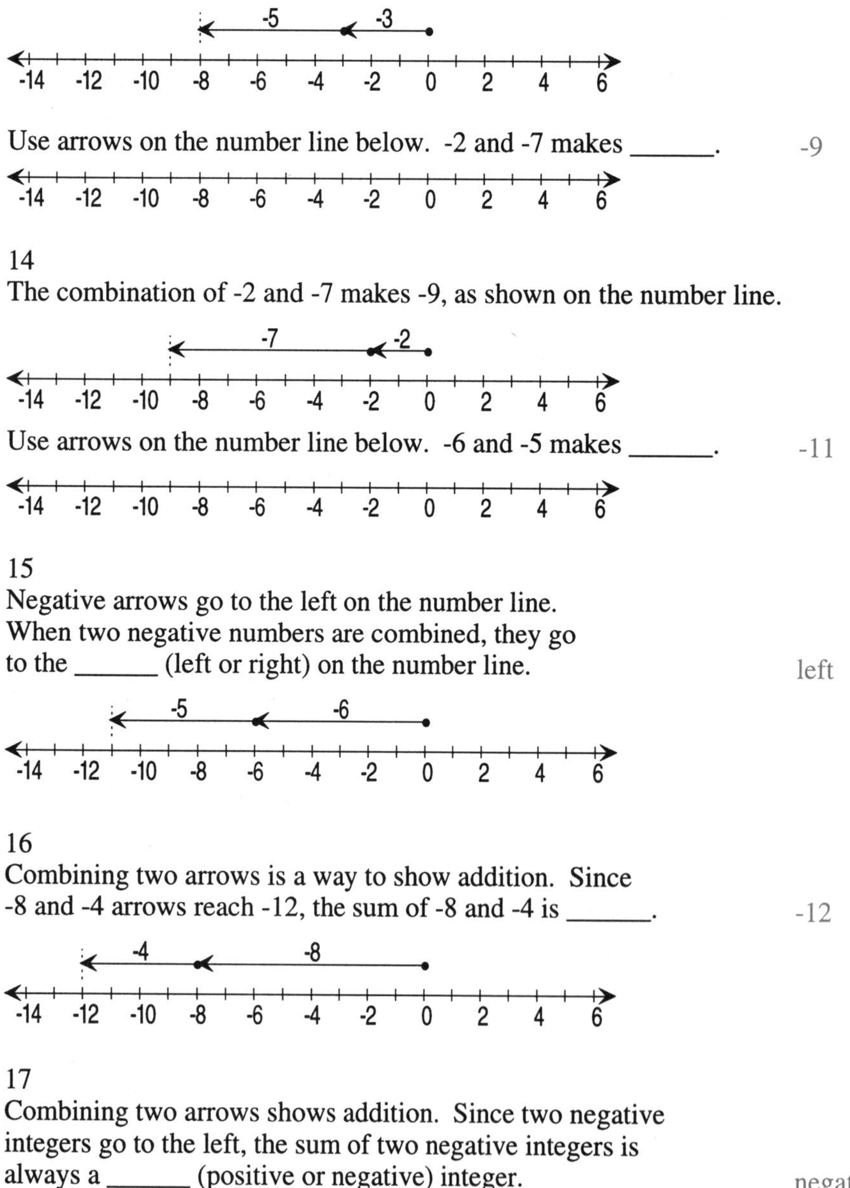

Use arrows on the number line below. -2 and -7 makes ______. -9

14
The combination of -2 and -7 makes -9, as shown on the number line.

Use arrows on the number line below. -6 and -5 makes ______. -11

15
Negative arrows go to the left on the number line. When two negative numbers are combined, they go to the ______ (left or right) on the number line. left

16
Combining two arrows is a way to show addition. Since -8 and -4 arrows reach -12, the sum of -8 and -4 is ______. -12

17
Combining two arrows shows addition. Since two negative integers go to the left, the sum of two negative integers is always a ______ (positive or negative) integer. negative

18
The sum of two negative integers is a negative integer.
-7 + (-2) = -9 and -4 + (-8) = ______ -12

19
-5 + (-7) = -12 and -9 + (-1) = ______ -10

20
-13 + (-4) = -17 and -2 + (-3) = ______ -5

21
-4 + (-2) = ______ -6

22
-7 + (-8) = ______ -15

The sum of two negative integers is always a negative integer.

-3 + (-8) = -11 -9 + (-7) = -16

23
When two negative integers are added, the sum is negative. -12 + (-8) = -20 and -17 + (-6) = ______ -23

24
-12 + (-2) = ______ -14

25
-15 + (-17) = ______ -32

26
-14 + (-35) = ______ -49

27
When two positive integers are added, the sum is positive. 12 + 7 = 19 and 14 + 8 = ______ 22

28
21 + 3 = ______ 24

29
When two negative integers are added, the sum is negative. -9 + (-8) = ______ -17

30

-19 + (-5) = ______

-24

31

To combine +7 and -5, the following arrow picture is used:

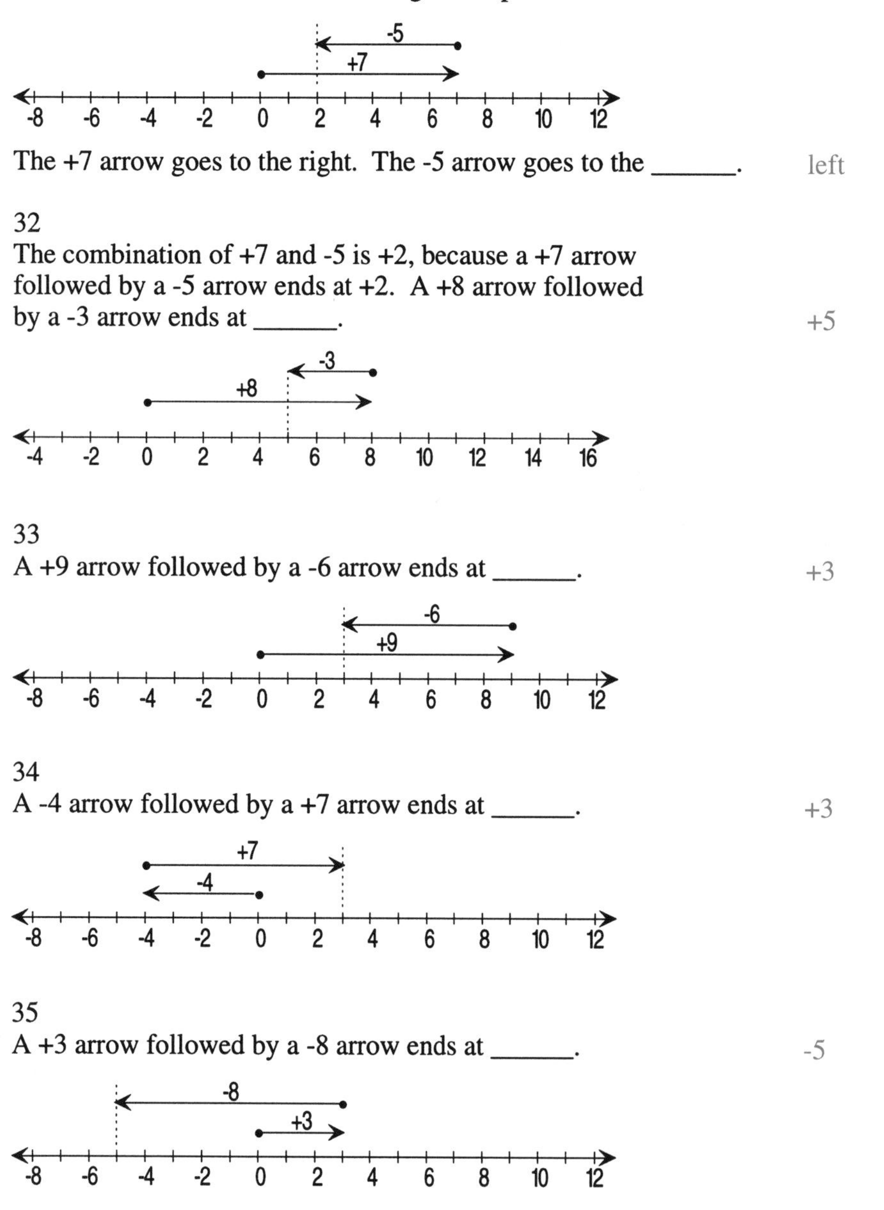

The +7 arrow goes to the right. The -5 arrow goes to the ______.

left

32

The combination of +7 and -5 is +2, because a +7 arrow followed by a -5 arrow ends at +2. A +8 arrow followed by a -3 arrow ends at ______.

+5

33

A +9 arrow followed by a -6 arrow ends at ______.

+3

34

A -4 arrow followed by a +7 arrow ends at ______.

+3

35

A +3 arrow followed by a -8 arrow ends at ______.

-5

36

Combining two arrows is **addition of integers**. Since -4 and +7 reach +3, the sum of -4 and +7 is ______.

+3

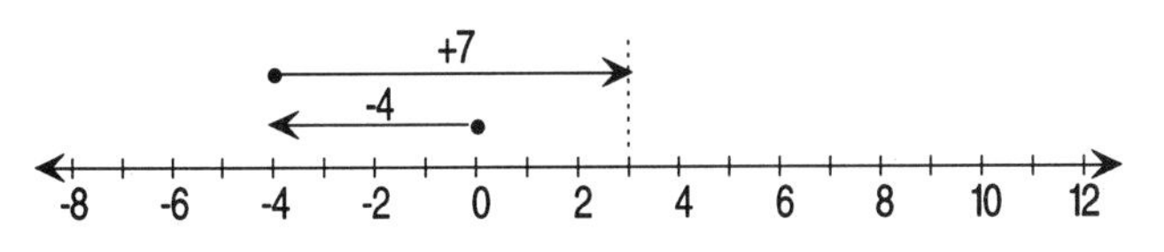

37

Combining two arrows is addition of integers. Since +10 and -12 reach -2, the sum of +10 and -12 is ______.

-2

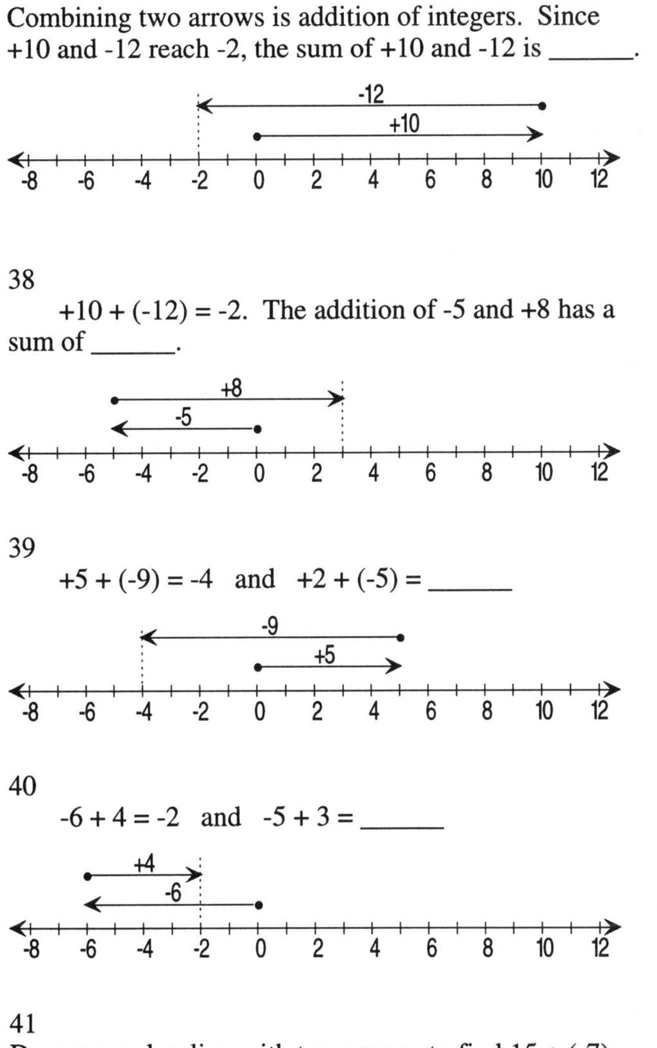

38

+10 + (-12) = -2. The addition of -5 and +8 has a sum of ______.

+3

39

+5 + (-9) = -4 and +2 + (-5) = ______

-3

40

-6 + 4 = -2 and -5 + 3 = ______

-2

41

Draw a number line with two arrows to find 15 + (-7).

+8 or 8

42

Draw a number line with two arrows to find -3 + 14.

+11 or 11

43
Draw a number line with two arrows to find -17 + 5. -12

44
Draw a number line with two arrows to find 9 + (-14). -5

45
Draw a number line with two arrows to find -2 + 8. +6 or 6

The sum of a positive and a negative integer may be positive, negative, or zero. The answer depends on which integer is farther from zero.

For example, in adding 7 and -3, the positive integer is farther from zero. Therefore, the sum is positive.

$$7 + (-3) = 4$$

In adding 5 and -9, the negative integer is farther from zero. Therefore, the sum is negative.

$$5 + (-9) = -4$$

In adding 6 and -6, the integers are the same distance from zero. Therefore, the sum is zero.

$$6 + (-6) = 0$$

46
8 + (-5) = ______ 3

47
13 + (-5) = ______ 8 or +8

48
-8 + 3 = ______ -5

49
-13 + 2 = ______ -11

50
15 + (-7) = ______ 8

51
12 + (-17) = ______ -5

52
-9 + 14 = ______ 5

53

-4 + 7 = ______ 3

54

-14 + 11 = ______ -3

55

4 + (-11) = ______ -7

56

When adding two positive integers, the sum is positive. 7 + 9 = 16 and 8 + 13 = ______ 21

57

7 + 17 = ______ 24

58

When adding two negative integers, the sum is negative. -7 + (-5) = -12 and -9 + (-8) = _____ -17

59

-15 + (-14) = ______ -29

60

-17 + (-45) = ______ -62

61

When adding a positive and a negative integer, the sign is determined by the longer arrow.

7 + (-3) = 4

7 has a ______ (longer or shorter) arrow than -3. longer

62

When adding a positive and a negative integer,the sign of the answer is determined by the longer arrow.

-16 + 9 = -7

-16 has a ______ (longer or shorter) arrow than 9. longer

63

-12 + 8 = ______ -4

At this time you must understand the following facts about adding integers:

1. If two integers are positive, their sum is positive.
2. If two integers are negative, their sum is negative.
3. If a positive integer is added to a negative integer, the sum may be positive, negative, or zero.

$8 + (-3) = 5$ $12 + (-19) = -7$ $4 + (-4) = 0$

64

$7 + 12 =$ ______ 19

65

$4 + 10 =$ ______ 14

66

$-6 + (-8) =$ ______ -14

67

$-21 + 6 =$ ______ -15

68

$17 + (-8) =$ ______ 9

69

$-7 + 0 =$ ______ -7

70

$-14 + (-1) =$ ______ -15

71

$0 + 0 =$ ______ 0

72

$9 + (-3) =$ ______ 6

73

$17 + 16 =$ ______ 33

74

$-4 + 7 =$ ______ 3

75

$-15 + 7 =$ ______ -8

76
7 + (-3) = ______ 4

77
-5 + (-4) = ______ -9

78
4 + (-19) = ______ -15

79
-9 + (-3) = ______ -12

80
-12 + 5 = ______ -7

81
0 + (-6) = ______ -6

82
-8 + 3 = ______ -5

83
-15 + (-6) = ______ -21

84
-15 + 3 = ______ -12

85
-7 + (-6) = ______ -13

86
8 + (-5) = ______ 3

87
-13 + 5 = ______ -8

88
7 + (-9) = ______ -2

89
-14 + (-12) = ______ -26

90
6 + (-5) = ______ 1

91
-19 + 4 = ______ -15

92

-21 + (-3) = ______ -24

93

-13 + (-7) = ______ -20

94

17 + (-17) = ______ 0

95

-8 + (-4) = ______ -12

FEEDBACK UNIT 2

This quiz reviews the preceding unit. Answers are at the back of the book.

1. 13 + 7 = ______
2. The sum of two positive integers is a ______ integer.
3. -5 + (-8) = ______
4. The sum of two negative integers is a ______ integer.
5. -2 + 6 = ______
6. 7 + (-7) = ______
7. -11 + (-4) = ______
8. 3 + (-11) = ______
9. -4 + (-19) = ______
10. -5 + 12 = ______
11. -15 + 15 = ______
12. 8 + (-3) = ______
13. 12 + (-19) = ______
14. -3 + (-11) = ______
15. 5 + (-7) = ______
16. -3 + (-19) = ______
17. 5 + (-17) = ______
18. -3 + (-7) = ______
19. 18 + (-13) = ______
20. 14 + (-25) = ______

UNIT 3: MULTIPLICATION OF INTEGERS

The following mathematical terms are crucial to an understanding of this unit.

Multiplication of Integers | Grouping

1
The multiplication of positive integers is the same as multiplying counting numbers.
3 • 5 = 15 and 7 • 4 = ______ | 28

2
The product of two positive integers is positive.
9 • 8 = ______ | 72

3
A positive integer multiplied by a positive integer always gives a positive integer. 6 • 3 = ______ | 18

4
Multiplication is a shortcut for special addition problems: 4 • 5 means 5 + 5 + 5 + 5 = 20.
3 • 6 means 6 + 6 + 6 = ______ | 18

5
2 • (-3) means -3 + (-3) = -6.
2 • (-5) means -5 + (-5) = ______ | -10

6
3 • (-6) = -6 + (-6) + (-6) = ______ | -18

7
3 • (-4) = -4 + (-4) + (-4) = ______ | -12

8
A positive integer multiplied by a negative integer always gives a negative integer.
4 • (-5) = -20 and 6 • (-7) = ______ | -42

9
The **product of a positive and a negative integer** is negative. 8 • (-4) = ______ | -32

10

$5 \cdot (-9) =$ ______ -45

11

$5 \cdot (-17) =$ ______ -85

Recall that the Commutative Law of Multiplication states that xy is equivalent to yx for all counting numbers. This important property also applies to the multiplication of integers. The order of two integers in multiplication can be reversed without changing the product.

12

The Commutative Law of Multiplication states that $5 \cdot 3 = 3 \cdot 5$. Therefore, $3 \cdot (-7) = -7 \cdot 3$. Since $3 \cdot (-7) = -21$, then $-7 \cdot 3 =$ ______. -21

13

The Commutative Law of Multiplication states that $3 \cdot (-9) = -9 \cdot 3$. Since $3 \cdot (-9) = -27$, then $-9 \cdot 3 =$ ______. -27

14

The Commutative Law of Multiplication states that $4 \cdot (-9) = -9 \cdot 4$. Since $4 \cdot (-9) = -36$, then $-9 \cdot 4 =$ ______. -36

15

Whenever a positive and a negative integer are multiplied, the product is negative. $-3 \cdot 12 = -36$ and $12 \cdot (-3) =$ ______. -36

16

The **product of a negative and a positive integer** is negative. $-3 \cdot 5 =$ ______ -15

17

$10 \cdot (-6) =$ ______ -60

18

$-6 \cdot 3 =$ ______ -18

19

$-7 \cdot 8 =$ ______ -56

20

$8 \cdot (-4) =$ ______ -32

21
$-12 \bullet 7 =$ ______ -84

22
$3 \bullet 0 = 0 + 0 + 0 = 0$ $2 \bullet 0 = 0 + 0 =$ ______ 0

23
Any integer **multiplied by zero** gives zero as the product.
$15 \bullet 0 = 0$, $-18 \bullet 0 = 0$, and $93 \bullet 0 =$ ______. 0

24
By the Commutative Law of Multiplication,
$3 \bullet 0 = 0 \bullet 3$. Since $7 \bullet 0 = 0$, then $0 \bullet 7 =$ ______. 0

25
$0 \bullet 5 =$ ______ 0

26
$0 \bullet (-9) =$ ______ 0

27
$0 \bullet 0 =$ ______ 0

The product of two positive integers is a positive integer.

The product of a positive integer and a negative integer is a negative integer.

The product of two negative integers is a positive integer.

The example below illustrates why an effort to maintain some other valuable properties forces the conclusion that $-5 \bullet -2$ must have a product of $+10$.

$-5 \bullet -2 = -5 \bullet -2 + 0$	$-5 \bullet -2$ must be equal to $-5 \bullet -2 + 0$
$= -5 \bullet -2 + (-10 + 10)$	$(-10 + 10)$ is substituted for 0
$= -5 \bullet -2 + (-5 \bullet 2 + 10)$	$-5 \bullet 2$ is substituted for -10
$= (-5 \bullet -2 + -5 \bullet 2) + 10$	An Associative grouping
$= -5(-2 + 2) + 10$	The Distributive property
$= -5(0) + 10$	0 is substituted for $-2 + 2$
$= 0 + 10$	$-5(0)$ is equal to 0
$= 10$	10 is equal to $0 + 10$

The series of equalities above shows that $-5 \bullet -2$ must be equal to $+10$. The reader is not responsible for reproducing this type of reasoning, but should now accept the fact that the following rule is reasonable and logical: The product of two negative integers is a positive integer.

28
The **product of two negative integers is a positive integer**. $-5 \cdot (-2) = +10$.
$-6 \cdot (-4) =$ ______ 24

29
A negative times a negative is a positive.
$-6 \cdot (-9) = +54$.
$-7 \cdot (-4) =$ ______ 28

30
The product of two negative integers is a positive integer. $-3 \cdot (-4) = 12$ and
$-9 \cdot (-5) =$ ______ 45

31
$-8 \cdot (-5) =$ ______ 40

32
$-12 \cdot (-10) =$ ______ 120

33
$-3 \cdot (-9) =$ ______ 27

34
$-7 \cdot (-3) =$ ______ 21

35
In multiplying two integers, if the signs are the same, the product is positive. If the signs are different, the product is negative. $6 \cdot (-5) =$ ______ -30

36
In multiplying two integers, if the signs are the same, the product is positive. If the signs are different, the product is negative. $-8 \cdot (-3) =$ ______ 24

37
$4 \cdot 8 =$ ______ 32

38
$-7 \cdot 1 =$ ______ -7

39
$-3 \cdot (-2) =$ ______ 6

40

0 • (-17) = ______ 0

41

-1 • 71 = ______ -71

42

-4 • (-6) = ______ 24

43

5 • 7 = ______ 35

44

3 • (-9) = ______ -27

45

To evaluate [5 • (-2)] • 3, the square brackets indicate that the first step is to multiply 5 and -2.

[5 • (-2)] • 3
-10 • 3
-30

Evaluate. [6 • (-2)] • 4 -48

Multiplication can be indicated by the multiplication dot or the use of parentheses. The examples below show the meaning of these two ways of indicating multiplication.

3 • 7 may be written as 3 • (7) or 3(7)
7 • (-5) may be written as 7(-5) or 7 • -5

46

Sometimes, parentheses indicate the first step in a multiplication evaluation. To evaluate (4 • 3) • -2, the following steps are used:

(4 • 3) • -2
12 • -2
-24

Evaluate. 4 • (3 • -2) -24

47

Recall that the Associative Law of Multiplication, as it applies to counting numbers, assures that (4 • 3) • 2 and 4 • (3 • 2) will have the same evaluation. Separately find the evaluations of 4 • (5 • -3) and (4 • 5) • -3. Both are -60

48
Do -3 • (7 • -5) and (-3 • 7) • -5 have the same evaluation?

Yes, both are 105

49
Do 6 • (-9 • 7) and (6 • -9) • 7 have the same evaluation?

Yes

The Associative Law of Multiplication applies to the integers and states that 5 • -3 • -7 can be evaluated as (5 • -3) • -7 or 5 • (-3 • -7). Whenever three integers are to be multiplied, the **grouping** will not affect the product.

50
To evaluate -3 • 7 • -5, either of the following may be used:

-3 • 7 • -5	or	-3 • 7 • -5
(-3 • 7) • -5		-3 • (7 • -5)
-21 • -5		-3 • -35
105		105

Evaluate. -4 • 3 • 5

-60

51
Evaluate. 2 • -7 • 3

-42

52
Evaluate. 4 • -7 • -2

56

53
Evaluate. 4 • (-2 • 9)

-72

54
Evaluate. 2 • 5 • -12

-120

55
Evaluate. 4 • 5 • -2

-40

56
Evaluate. 5 • (-2 • 3)

-30

57
Evaluate. -2 • 4 • -3

24

58
Evaluate. 8 • -5 • -2

80

59
Evaluate. (-2 • -4) • -3

-24

60
Evaluate. -3 • -3

9

61
Evaluate. 2 • 2 • 2

8

62
Evaluate. -1 • -1 • -1

-1

FEEDBACK UNIT 3

This quiz reviews the preceding unit. Answers are at the back of the book.

1. 8 • 3 = ______
2. The product of two positive integers is a ______ integer.
3. 4 • -5 = ______
4. The product of a positive integer and a negative integer is a ______ integer.
5. -3 • -7 = ______
6. The product of two negative integers is a ______ integer.
7. -9 • -6 = ______
8. 7 • 0 = ______
9. -14 • 2 = ______
10. -3 • -12 = ______
11. 4 • -1 = ______
12. 0 • -5 = ______
13. 5 • -2 • 6 = ______
14. -2 • -2 • -2 = ______
15. -3 • 5 • -8 = ______
16. 3(4 • -2) = ______
17. -2 • -2 • -5 = ______
18. (5 • -3) • -2 = ______
19. -3 • -2 • -3 = ______
20. -5 • 2 • 3 = ______

UNIT 4: SUBTRACTION OF INTEGERS AND EVALUATING NUMERICAL EXPRESSIONS

The following mathematical terms are crucial to an understanding of this unit.

Minus sign Subtraction of integers

1

Whenever a **minus sign** is placed between two integers, it means to add the opposite of the second integer to the first integer. For example, 7 – 3 is the same as 7 + (-3). Is 8 – 4 the same as 8 + (-4)?

Yes

2

9 – 11 means 9 plus -11 or 9 + (-11).
4 – 7 means 4 plus -7 or ______

4 + (-7)

3

-4 – 6 means -4 plus -6 or -4 + (-6).
8 – 9 means 8 plus -9 or ______

8 + (-9)

4

-8 – (-9) means -8 plus 9 or -8 + 9.
-3 – (-6) means -3 plus 6 or ______

-3 + 6

5

The minus sign should always be read as part of the number to its right. In the expression 6 – 5 the minus is read as part of the 5. In the expression 4 – 8, the minus is read as part of the ______.

8

6

5 – 3 is read as 5 plus -3.
8 – 4 is read as 8 plus ______

-4

7

10 – 7 is read as 10 plus -7.
9 – 5 is read as 9 plus ______

-5

8
The expression 8 – 1 involves the integers 8 and -1.

8 – 1 = (8) (–1) = 8 + (-1)

The expression 9 – 7 involves the integers 9 and ______. -7

9
In the expression -5 + 4 or (-5)(+ 4) the integers are -5 and 4. In the expression -7 – 3 or (-7) (– 3) the integers are -7 and ______. -3

10
In the expression 6 – 10 or (6) (–10) the integers are 6 and ______. -10

A double negative such as –(-3) is positive, because –(-3) is read as: "the opposite of -3." Consequently, –(-3) is +3, and the problem 7 – (-3) can be read as:

(7)(–(-3)) or (7)(+ 3)

+(-3) is -3, and the problem 8 + (-5) can be read as:

(8)(+(-5)) or (8) (– 5)

11
–(-3) is equal to 3. 8 – (-3) involves the integers 8 and 3.
5 – (-7) involves the integers 5 and ______. 7

12
-8 + 7 involves the integers -8 and ______. 7

13
5 – 3 involves the integers 5 and ______. -3

14
7 – (-2) involves the integers 7 and ______. 2

15
3 – (-8) involves the integers 3 and ______. 8

16
-5 + 9 involves the integers -5 and ______. 9

The minus sign is associated with the operation **subtraction**. In this unit, we have shown that the use of the minus sign should not be read as "take away." The minus sign is read as part of the number to its right, and the signs of the integers are used to evaluate the expressions.

17
To evaluate the expression 5 – 4 + 7, the plus and minus signs should be read as part of the numbers.
(5)(– 4)(+ 7)
is an expression involving 5, -4, and ______.

7

18
To evaluate -6 – 9 + 2, read each number with the sign preceding it.
(-6)(– 9)(+ 2)
involves the numbers -6, -9, and ______.

2

19
The numbers in the expression 6 – 4 – 9 are read as 6, -4, and ______.

-9

20
To evaluate 4 – 7 – 8, read each number with the sign preceding it.
(4)(– 7)(– 8)
involves the numbers 4, -7, and ______.

-8

21
To evaluate -5 + 7 – 9, read the numbers as -5, 7, and ______.

-9

22
To evaluate 2 – 6 + 5, read the numbers as 2, -6, and ______.

5

23
To evaluate 7 – 3 – 9, think of the numbers as 7, -3, and ______.

-9

24

The evaluation of -3 – 9 is shown below.

-3 – 9

(-3)(– 9)

-12

The sum of -3 and -9 is -12. Evaluate -6 – 7.

-13

25

The evaluation of 6 – 11 is shown below.

6 – 11

(6)(– 11)

-5

The sum of 6 and -11 is -5. Evaluate 4 – 10.

-6

26

The evaluation of -5 – 1 is shown below.

-5 – 1

(-5)(– 1)

-6

The sum of -5 and -1 is -6. Evaluate -7 – 4.

-11

27

To evaluate 5 – 8 means to add 5 and -8.
Evaluate 5 – 8.

-3

28

Evaluate. -11 – 6

-17

29

Evaluate. 10 – 3

7

30

Evaluate. 9 – 14

-5

31

"–(-5)" is read as: the opposite of -5. As in multiplication, two negatives make a positive. "–(-5)" is equal to +5. "–(-7)" is read as: the opposite of -7 or ______.

7

32

"–(-6)" is equal to +6. "–(-8)" is read as: the opposite of -8 or ______.

8

33

Read "–(-9)" as +9. Read "–(-11)" as: the opposite of -11 or ______.

11

34

To evaluate 14 – (-3), the following steps are used:

14 – (-3)

(14) (– (-3))

14 + 3

17

Evaluate. 15 – (-6)

21

35

To evaluate -4 – (-7), the following steps are used:

-4 – (-7)

(-4) (– (-7))

-4 + 7

3

Evaluate. -6 – (-11)

5

36

7 – 12 = (7)(– 12) = -5

14 – 20 = (14)(– 20) = ______

-6

37

-4 – 7 = (-4)(– 7) = -11

-6 – 17 = ______

-23

38

-7 – (-2) = (-7)(+ 2) = -5

-13 – (-9) = ______

-4

39

14 – (-3) = (14)(+ 3) = 17

11 – (-5) = ______

16

40

The evaluation of 5 – 3 is 2 because 5 – 3 means to add (5) and (-3). Evaluate 8 – 2.

6

41
Evaluate. 15 – 7

8

42
Evaluate. -12 – (-4)

-8

43
Evaluate. 9 – 12

-3

44
Evaluate 10 – (-5) by adding 10 and 5.

15

45
Evaluate. -4 – (-10)

6

46
The evaluation of 14 – 9 is 5. Evaluate 13 – 10.

3

47
Evaluate. 7 – 3

4

48
Evaluate. 5 – (-3)

8

49
Evaluate. 9 – 0

9

50
Evaluate. 17 – (-4)

21

51
Evaluate. 7 – 6

1

52
Evaluate. 14 – (-17)

31

53
Evaluate. -9 – 3

-12

54
Evaluate. 7 – (-7)

14

55
Evaluate. -17 – 13

-30

56
Evaluate. 15 – 15

0

57
Evaluate. 0 – 2

-2

58
Evaluate. 27 – 9

18

59
Evaluate. 8 – 3

5

60
Evaluate. -21 – 7

-28

61
Evaluate. -9 – (-4)

-5

62
Evaluate. 9 – 13

-4

63
Evaluate. -8 – (-10)

2

To do the problem 10 – 4 – 2, read each of the numbers with the sign preceding it.
(10) (– 4) (– 2) involves adding the integers 10, -4, -2.

10 – 4 – 2 = (10) (– 4) (– 2) = 4

64
To evaluate -7 – 8 + 7, the following steps are used:

(-7)(–8)(+7)

-15 + 7

-8

Evaluate. -6 – 7 + 12

-1

65
Evaluate. -4 – 3 + 8

1

66
Evaluate. 2 – 5 – 1

-4

67
Evaluate. 14 – 3 – 33

-22

68
To evaluate -4 – 7 – (-12), the following steps are used:

-4 – 7 – (-12)

(-4) (– 7) (+ 12)

-11 + 12

1

Evaluate. -7 – 10 – (-3)

-14

69
Evaluate. 5 – 8 – 3

-6

70
Evaluate. 12 – (-3) – 5

10

71
Evaluate. -5 – 4 – 8

-17

72
Evaluate. 4 – 3 – 8

-7

73
Evaluate. -7 – 6 – 5

-18

74
To evaluate 8 + (-3) – (-4), the following steps are used:

8 + (-3) – (-4)

8 – 3 + 4

5 + 4

9

Evaluate. -3 + (-4) – (-3)

-4

75
Evaluate. 3 – (-7) – (-5) – 6

9

76
Evaluate. -12 + 9 – (-12) + 4

13

77
Evaluate. 6 – (-17) – 8

15

78
Evaluate. -24 – (-13) – 7

-18

79
Evaluate. 5 + 9 + (-10) – (-30) 34

80
Evaluate. 5 – 7 + 3 – 9 -8

81
Evaluate. -4 + 15 – 12 + 17 16

82
Evaluate. -15 – (-15) 0

83
Evaluate. -9 + (-6) + 8 – (-3) -4

84
Evaluate. 14 – 9 – 5 + 12 12

85
Evaluate. -8 – 17 + (-6) – (-31) 0

FEEDBACK UNIT 4

10/25/08

This quiz reviews the preceding unit. Answers are at the back of the book.

1. Evaluate. -9 – 3
2. Evaluate. 14 – 5
3. Evaluate. 4 – (-7)
4. Evaluate. 2 – 5 – 3
5. Evaluate. -3 – (-2) + 4
6. Evaluate. 4 – 7 – (-7)
7. Evaluate. -3 – 6 + 4
8. Evaluate. 7 – 7 – (-2)
9. Evaluate. 2 – 3 – 6
10. Evaluate. -5 – 4 – (-2)
11. Evaluate. -4 + (-5) + (-7)
12. Evaluate. 5 – (-9) – (-5)
13. Evaluate. -7 + 4 – 9 – 14
14. Evaluate. -4 – 8 – 15 – 19
15. Evaluate. 3 – 15 + 18 – (-3)

UNIT 5: EVALUATING NUMERICAL EXPRESSIONS INVOLVING MULTIPLICATION AND EXPONENTS

The following mathematical terms are crucial to an understanding of this unit.

Factors	Exponent
Second power	Squared
Third power	Cubed
Base	Order of operations

1
In the multiplication expression 2 • -5 • 3 the numbers 2, -5, and 3 are called **factors**. The factors of 5 • -3 are 5 and ______.

-3

2
The factors of 3 • -2 • 4 are 3, -2, and ______.

4

3
The factors of 5 • 6 • 2 are 5, 6, and ______.

2

4
The factors of -6 • 5 are -6 and ______.

5

5
A number in a multiplication expression is called a factor. For the numerical expression 7 • 8, the numbers 7 and 8 are called ______.

factors

6
The factors of the multiplication expression 6 • -7 • 4 are ______, ______, and ______.

6, -7, 4

7
Evaluate. 3 • 3 • 3

27

8
Evaluate. 8 • 8

64

9
The numerical expression with three factors of 7 can be written as $7 \cdot 7 \cdot 7$ or 7^3. The small three is an **exponent**. What is the exponent of 6^2?

2

10
6^2 which means $6 \cdot 6$ is read "6 to the **second power** or 6 **squared**." Write the expression for 9 squared.

9^2

11
7^3 which means $7 \cdot 7 \cdot 7$ is read "7 to the **third power** or 7 **cubed**." Write the expression for 5 cubed.

5^3

12
The fourth power of 9 can be written as $9 \cdot 9 \cdot 9 \cdot 9$ or 9^4. Write the expression for the fourth power of 3.

3^4

13
Use an exponent to write an expression for $10 \cdot 10 \cdot 10$.

10^3

14
Use an exponent to write an expression for $7 \cdot 7 \cdot 7 \cdot 7 \cdot 7 \cdot 7 \cdot 7 \cdot 7$.

7^8

15
To write $-3 \cdot -3 \cdot -3 \cdot -3 \cdot -3$ with an exponent, parentheses must be used.

$-3 \cdot -3 \cdot -3 \cdot -3 \cdot -3 = (-3)^5$

Use an exponent to write an expression for $-6 \cdot -6 \cdot -6 \cdot -6 \cdot -6$.

$(-6)^5$, parentheses must be used

16
Use an exponent to write an expression for $10 \cdot 10 \cdot 10 \cdot 10$.

10^4

17
Use an exponent to write an expression for $-5 \cdot -5 \cdot -5$.

$(-5)^3$, parentheses must be used

18
In the expression 2^3 the 2 is the **base** and the 3 is the exponent. What is the base of 5^4?

5

19
If the base of an exponent is negative, parentheses will enclose the negative sign. Otherwise, the base is positive. In the expression, -4^5 the base is ____.

+4

20
-2^3 means $-1 \cdot 2^3$. Notice that the base is +2. When the negative sign is not enclosed in parentheses the base is ___________.

positive

21
Be careful in reading -4^5. It means $-1 \cdot 4^5$. Similarly, -7^3 means ______.

$-1 \cdot 7^3$, base is 7

22
If the base of an exponent is negative, the negative sign will be inside parentheses. The base of $(-3)^4$ is -3 and $(-3)^4$ is equal to ______.

$-3 \cdot -3 \cdot -3 \cdot -3 = 81$

23
9^3 is equal to ______.

$9 \cdot 9 \cdot 9 = 729$

24
If a negative base is used its sign must be enclosed by parentheses. Otherwise, the base is positive. The base of -2^5 is ______.

+2, $-2^5 = -1 \cdot 2^5$

25
To evaluate 5^3, the following steps are used:
$5^3 = 5 \cdot 5 \cdot 5 = 125$
Evaluate 4^3.

64

26
Complete the evaluation of 7^3.
$7^3 = 7 \cdot 7 \cdot 7 =$ ______

343

27
Complete the evaluation of 1^5.
$1^5 = 1 \cdot 1 \cdot 1 \cdot 1 \cdot 1 =$ ______

1

28
Complete the evaluation of -2^4.
$-2^4 = -1 \cdot 2^4 = -1 \cdot (2 \cdot 2 \cdot 2 \cdot 2) =$ ____

-16

29
Complete the evaluation of $(-2)^4$.
$(-2)^4 = -2 \bullet -2 \bullet -2 \bullet -2 =$ _______

16

30
Evaluate. $(-9)^2$

81

31
Evaluate. -9^2

$-1 \bullet 9^2 = -81$

32
Evaluate. 6^2

36

33
To evaluate $-5(2 - 3)$, first perform the operation in parentheses.

$-5(2 - 3)$
$-5(-1)$
5

Evaluate. $4(6 - 4)$

8

34
Evaluate. $-2(8 - 11)$

6

35
Evaluate. $-6(-2 - 4)$

36

36
To evaluate $8 + (3 \bullet - 4)$, the following steps are used:

$8 + (3 \bullet - 4)$
$8 + (-12)$
$8 - 12$
-4

Evaluate. $9 + (2 \bullet -7)$

-5

37
Evaluate. $2 + (-3 \bullet 7)$

-19

38
Evaluate. $-4(-5 - 2)$

28

39
Evaluate. $6 - 5^2$

$6 - 25 = -19$

40
Evaluate. $-3 + (-5 \bullet -3)$

12

In evaluating a numerical expression that does not contain parentheses, all multiplication is done before any addition. This is the **order of operation**: If grouping symbols appear in an expression with more than one operation, they determine the order in which the operations are to be performed. If no grouping symbols are present, all multiplication precedes any addition.

4 • -2 + 3	7 + 4 • -5
-8 + 3	7 – 20
-5	-13

41
Complete the evaluation.
5 • -3 + 2 = -15 + 2 = ______

-13

42
Evaluate. -7 • 2 + (-5)

-19

43
Complete the evaluation. 7 + 8 • -3 = ______

7 – 24 = -17

44
Evaluate. -3 + 6 • 4

21

45
Evaluate. -3 • -4 – 5

7

46
Evaluate. $-7 + (-2)^3$

-7 + (-8) = -15

47
To evaluate 5 • -2 – 3, the following steps are used:

5 • -2 – 3
-10 – 3
-13

Evaluate. 7 • -3 – 6

-27

48
Complete the evaluation. 4 – 2 • 5 = ______

4 – 10 = -6

49
Evaluate. -9 • -5 + (-3)

42

50
Evaluate. 2 • -7 + (-4)

-18

51
Evaluate. $-7 - 4 \cdot -2$ 1

52
Evaluate. $-4 \cdot 5 - (-3)$ -17

53
Evaluate. $-3 \cdot 2 + (-5)^2$ $-6 + 25 = 19$

54
Evaluate. $3 - 3 \cdot 2$ -3

FEEDBACK UNIT 5

This quiz reviews the preceding unit. Answers are at the back of the book.

1. $-15 \cdot -4 =$ ______
2. $8 \cdot -9 =$ ______
3. $-5(-3 - 4) =$ ______
4. $[3 - (-4)] - (-6) =$ ______
5. $-2 + 6 \cdot 3 =$ ______
6. $(8 - 4) \cdot -3 =$ ______
7. $-9 \cdot -8 + (-3) =$ ______
8. $3^4 =$ ______
9. $(-2)^3 =$ ______
10. $1^5 =$ ______
11. $(-5)^2 =$ ______
12. $-5^2 =$ ______

UNIT 6: PRIME AND COMPOSITE NUMBERS

The following mathematical terms are crucial to an understanding of this unit.

Pair of factors	Positive integer factors
Prime number	Composite numbers

1

6 and 4 are factors of 24 because 6 • 4 = 24.
Are 8 and 3 also factors of 24? — Yes

2

7 and 3 are factors of 21 because 7 • 3 = 21.
Factors of 14 are 2 and ______. — 7

3

-4 and 9 is a **pair of factors** of -36 because -4 • 9 = -36. Two factors of -26 are -2 and ______. — 13

4

A pair of factors of -20 is 5 and ______. — -4

5

A pair of factors of 30 is -3 and ______. — -10

6

A pair of factors of 15 is 15 and ______. — 1

7

A pair of factors of -42 is 6 and ______. — -7

8

A pair of factors of 35 is -5 and ______. — -7

9

The set of all **positive integer factors** of 10 is {1, 2, 5, 10} because 1 • 10 = 10 and 2 • 5 = 10.
What is the set of all positive integer factors of 15? — {1, 3, 5, 15}

10

Write the set of all positive integer factors of 6. — {1, 2, 3, 6}

11
Write the set of all positive integer factors of 7.

{1, 7}

12
The number 7 has two elements in its set of positive integer factors, {1, 7}. Find the set of positive integer factors of 8 and state the number of elements in the set.

{1, 2, 4, 8}, 4

13
Write the set of all positive integer factors of 12.

{1, 2, 3, 4, 6, 12}

14
{1, 2, 3, 4, 6, 12} has six elements. This means there are ______ (how many) positive integers that are factors of 12.

6

15
Write the set of all positive integer factors of 5.

{1, 5}

16
{1, 5} has two elements. How many positive integers are factors of 5?

2

17
Write the set of all positive integer factors of 9.

{1, 3, 9}
(It is not necessary to include 3 twice in the set.)

18
{1, 3, 9} is the set of all positive integer factors of 9. How many positive integers are factors of 9?

3

19
Write the set of all positive integer factors of 4.

{1, 2, 4}

20
{1, 2, 4} is the set of all positive integer factors of 4. How many positive integers are factors of 4?

3

21
5 is a **prime number** because its set of factors has exactly two elements, {1, 5}. 7 is also a prime number because its set of factors has exactly ______ elements, {1, 7}.

2

22

11 is a prime number because it has exactly two positive integer factors. The positive integer factors of 11 are 1 and ______.

11

23

Any integer that has only itself and 1 as positive integer factors is called a prime number. Is 13 a prime number?

Yes

24

6 is not a prime number because its set of factors, {1, 2, 3, 6}, has more than two elements. 8 is not a prime number because its set of factors, {1, 2, 4, 8}, has ______ (more, less) than two elements.

more

25

2 is a prime number because 2 and 1 are the only positive integer factors of 2. Is 3 a prime number?

Yes

26

5 is a prime number because 5 and 1 are the only positive integer factors of 5. Is 7 a prime number?

Yes

27

6 is not a prime number because it has more than one pair of positive integer factors; $1 \cdot 6 = 6$ and $2 \cdot 3 = 6$. $1 \cdot 10 = 10$ and $2 \cdot 5 = 10$. Is 10 a prime number?

No

28

$1 \cdot 15 = 15$ and $3 \cdot 5 = 15$. Is 15 a prime number?

No

29

10 and 15 are not prime numbers because they have more than one pair of positive integer factors. 10 and 15 are **composite numbers**. $1 \cdot 14 = 14$ and $2 \cdot 7 = 14$. Is 14 a composite number?

Yes

30

11 is a prime number because it has only one pair of positive integer factors; 1 • 11 = 11. Is 17 a prime number?

Yes

31

18 is a composite number because it has more than one pair of of positive integer factors; 1 • 18 = 18, 2 • 9 = 18, and 3 • 6 = 18. Is 27 a composite number?

Yes

32

13 has only one pair of positive integer factors; 1 • 13 = 13. Is 13 a prime number?

Yes

33

1 • 8 = 8 and 2 • 4 = 8. 8 is a ______ (prime, composite) number.

composite

34

1 • 19 = 19. 19 is a ______ (prime, composite) number.

prime

35

1 • 20 = 20, 2 • 10 = 20, and 4 • 5 = 20. 20 is a ______ (prime, composite) number.

composite

36

24 is a ______ (prime, composite) number.

composite

37

23 is a ______ number.

prime

38

31 is a ______ number.

prime

39

34 is a ______ number.

composite

40

81 is a ______ number.

composite

41

38 is a ______ number.

composite

42

36 is a ______ number.

composite

43

37 is a ______ number. prime

44

47 is a ______ number. prime

45

The first four prime numbers are 2, 3, 5, and 7.
List all the prime numbers between 10 and 20. 11, 13, 17, 19

46

List all the prime numbers between 20 and 30. 23, 29

47

List all the prime numbers between 30 and 40. 31, 37

48

List all the prime numbers between 40 and 50. 41, 43, 47

FEEDBACK UNIT 6

This quiz reviews the preceding unit. Answers are at the back of the book.

1. Which of the following integers are primes?
   2, 7, 10, 15, 17, 20, 24, 30, 31
2. Which of the following integers are composites?
   3, 4, 5, 10, 13, 17, 19, 25, 30
3. List the first five prime numbers.
4. List the prime numbers between 50 and 60.
5. List the prime numbers between 15 and 25.
6. List the composite numbers between 31 and 38.
7. List the prime numbers between 55 and 65.
8. List the prime numbers between 80 and 90.

UNIT 7: APPLICATIONS

In this Applications Section, the format of the text has been altered. Answers for the problems appear beneath them rather than in the right-hand column. Your studying emphasis should be on learning the best procedures to follow with word problems. For that reason, once the procedure is learned a calculator may be used to complete the answer.

1

Many real-world relationships have been translated into formulas. For example, a basic understanding for every business is the relationship between costs, profits, and sales. This relationship is often shown as a formula.

C	+	P	=	S
Costs	plus	Profits	equals	Sales

Use the formula to write an equation for the following situation: Apex Company last month had sales of $18,000 and costs of $16,900.

Answer: For this problem, rewrite the formula, C + P = S, replacing S by 18,000 and C by 16,900.

C	+	P	=	S
16,900	+	P	=	18,000

The equation has 1,100 as its solution. Therefore, the profit (P) was $1,100.

2

Use the formula, C + P = S, to write an equation related to the following situation. Mel's Bookstore had a profit last year of $7,500 on sales of $83,000.

Answer: For this problem, rewrite the formula, C + P = S, replacing S by 83,000 and P by 7,500.

C	+	P	=	S
C	+	7,500	=	83,000

The equation has 75,500 as its solution. Therefore, the costs (C) were $75,500.

3

Another real-world problem that has been translated into a formula gives the relationship between a square's perimeter and the length of one of its sides. The formula is P = 4s.

P	=	4	•	s
Square's Perimeter	equals	4	times	length of a side

Use the formula to write an equation for the following situation:
A gardener wishes to use 48 feet of fencing to enclose a square plot.
What would be the length of one side of the plot?

Answer: For this problem, rewrite the formula, P = 4s, replacing P by 48.

P	=	4	•	s
48	=	4	•	s

The equation has 12 as its solution. Therefore, the length of the plot's side is 12 feet.

4

The value of a collection of objects where each member of the collection has the same price or cost is written as the formula V = ne

V	=	n	•	e
Collection's Value	equals	number of items	times	price of each item

Use the formula to write an equation for the following situation: A stack of quarters has a value of $9.75 and the number of quarters needs to be determined.

Answer: For this problem, rewrite the formula, V = ne, replacing V by 9.75 and e by .25 as shown below.

V	=	n	•	e
9.75	=	n	•	.25

The equation has 39 as its solution. Therefore, there are 39 quarters in the stack.

5

The formula D = RT is often called the Distance Formula and shows the relationship between distance traveled, uniform speed or rate, and time.

D	=	R	•	T
Distance	equals	Speed per time unit	times	Number of time units

Use the formula to write an equation for the following situation: On a transcontinental flight the pilot was able to cruise for 3 hours at 650 miles per hour. Find the distance traveled during that time.

Answer: For this problem, rewrite the formula, D = RT, replacing R by 650 and T by 3 as shown below.

D	=	R	•	T
D	=	650	•	3

The equation has 1950 as its solution. Therefore, the plane traveled 1,950 miles.

6

One of the percent formulas can be used to write an equation for any of the percent problems. For example, the formula for part is p = rb.

p	=	r	•	b
part	equals	rate	times	base

Use the formula to write an equation for the following situation: 54% of the voters in a local election were women. If 27,000 women voted, what was the total vote?

Answer: For this problem, rewrite the formula, p = rb, replacing r by .54 and p by 27,000 as shown below.

p	=	r	•	b
27,000	equals	.54	times	base

The equation has 50,000 as its solution. Therefore, 50,000 people voted.

FEEDBACK UNIT 7 FOR APPLICATIONS

1. Use the formula C + P = S to write an equation for the following situation. Apex Cleaners had costs last month of $43,000 and sales of $51,000. The owner would like to be able to determine his profit.

2. Use the formula P = 4s to write an equation for the following situation. A square table cloth is 6 feet long on each side? How many feet of fringe material would be needed for the perimeter of the table cloth?

3. Use the formula V = ne to write an equation for the following situation. The Art Theater sold 243 tickets last night for $3.25 each. What was the total income from those tickets?

4. Use the formula D = RT to write an equation for the following situation. Art Smith's race car can average 83 miles per hour on the Denver Course. How long will it take Art to travel 415 miles on that course?

5. Use the formula p = rb to write an equation for the following situation. Acme Publishing is revising one of its popular books. 74 pages are to be altered out of the 576 pages in the book. What percent of the pages are being revised?

SUMMARY FOR THE ARITHMETIC OF THE INTEGERS

The following mathematical terms are crucial to an understanding of this chapter.

Opposite	Negative	Negative integers
Integers	Set of integers	Positive integers
Number line	Addition of integers	Multiplication of integers
Grouping	Minus sign	Subtraction of integers
Factors	Exponent	Second power
Squared	Third power	Cubed
Base	Order of operations	Positive integer factors
Pair of factors	Prime number	Composite numbers

The set of integers consists of the set of counting numbers, zero, and the opposite of each counting number:

$$\{\ldots, -3, -2, -1, 0, 1, 2, 3, \ldots\}.$$

Every integer has an opposite. -13 is the opposite of 13, and 24 is the opposite of -24. The sum of an integer and its opposite is zero.

Addition and multiplication of integers was presented. Addition was shown by use of the number line, and multiplication was presented by stating the rules for the multiplication of signed numbers.

The product of two positive integers is positive.
The product of two negative integers is positive.
The product of a positive and a negative integer is negative.

Evaluating expressions that have two operations (multiplication and addition) was presented, together with the order of operations to complete evaluations correctly. Unless parentheses or other grouping symbols indicate otherwise, all multiplication is to be completed before the addition.

Prime and composite numbers were presented in Unit 6. A prime number is any integer greater than one that has only itself and 1 as factors. For example, 17 is a prime number because it only has factors of 1 and 17.

15 is a composite number because it has more than itself and 1 as factors. 15 has factors of 1, 3, 5, 15. Every composite number has more than two factors.

The counting number 1 is neither prime nor composite. All other counting numbers are prime or composite.

CHAPTER 4 MASTERY TEST

The following questions test the objectives of Chapter 4. Answers are at the back of the book. The number in parentheses which follows each problem indicates the unit in which it can be learned.

1. What is the opposite of 63? (1)
2. Solve. $x + 8 = 0$ (1)
3. Evaluate. $-9 + (-4)$ (2)
4. Evaluate. $4 - (-5)$ (2)
5. Evaluate. $-8 - 7$ (2)
6. Evaluate. $3 - 9$ (2)
7. Evaluate. $-6 \bullet 5$ (3)
8. Evaluate. $-8 \bullet -3$ (3)
9. Evaluate. $4 + 7 \bullet -3$ (4)
10. Evaluate. $-6 - 5 \bullet 7$ (4)
11. Evaluate. $-4 - 8 - 7$ (4)
12. Evaluate. $9 - (-3) + (-6)$ (4)
13. Evaluate. $3 - 7 - 2$ (4)
14. Evaluate. $-5(4 \bullet -2)$ (5)
15. Evaluate. $-3(-4) \bullet -2$ (5)
16. Evaluate. -5^3 (5)
17. Evaluate. $(-3)^4$ (5)
18. Evaluate. 2^5 (5)
19. List the prime numbers between 17 and 25. (6)
20. List the composite numbers between 8 and 15. (6)
21. Use the formula, $C + P = S$, which relates costs, profits, and sales to write an equation for the following situation. Della's Cookies had a profit last year of \$3,600 on sales of \$15,800. What were last year's costs? (7)
22. Use the formula, $D = RT$, which relates distance, speed, and time to write an equation for the following situation. Carl drives the interstate highway at a steady rate for 4 hours and travels 252 miles in that time. What was Carl's rate of speed? (7)

Chapter 5 Objectives

The following problems illustrate the objectives of this chapter. At this time you are not expected to know how to do these problems. However, if all of these problems are thoroughly understood, proceed directly to the Chapter 5 Mastery Test. The number in parentheses which follows each problem indicates the unit in which it can be learned.

Evaluate each of the following expressions.

1. $3x - 4$, when $x = 8$ (1)
2. $ab + 6$, when $a = -2$ and $b = 7$ (1)
3. $x^3 \cdot x$, when $x = -3$ (1)

Complete the following statements.

4. By the Associative Law of Addition, $9 + (3 + x) =$ ______ (2)
5. By the Commutative Law of Addition, $7x + 3 =$ ______ (2)

Simplify each of the following expressions.

6. $7 - 5x - 3$ (2)
7. $9x + 3 - 3x$ (2)
8. $5 \cdot (-7x)$ (3)
9. $(-6x) \cdot 5$ (3)
10. $x^5 \cdot x$ (4)
11. $x \cdot x$ (4)
12. $x^3 \cdot x^3$ (4)

Complete the following statements.

13. By the Commutative Law of Multiplication, $7x =$ ______ (3)
14. By the Associative Law of Multiplication, $-3(5 \cdot 9) =$ ______ (3)

Simplify each of the following expressions.

15. $9 - 3x + 3x - 7$ (5)
16. $-3(8 - 3x)$ (5)
17. $-4x - (5x + 1)$ (6)
18. $-(2x - 3) + (3x - 1)$ (6)
19. $3(2x - 5) - (x - 9)$ (6)
20. $8(5x - 7) - (5x - 2)$ (6)

CHAPTER 5
THE ALGEBRA OF THE INTEGERS

UNIT 1: EVALUATING NUMERICAL EXPRESSIONS

The following mathematical terms are crucial to an understanding of this unit.

Open expression
Numerical expression
Evaluation

1
If x is replaced by 3 in the **open expression** $2x + 5$, it becomes the **numerical expression** $2(3) + 5$ and has an **evaluation** of ______.

$6 + 5 = 11$

2
When 5 is used as a replacement in the open expression $-2 + 3x$, it is best to enclose it in parentheses. $-2 + 3(5) =$ ______.

$-2 + 15 = 13$

3
Use parentheses to replace x by -2 in the open expression $4 + 3x$. What is the evaluation of the numerical expression?

$4 + 3(-2) = -2$

4
Evaluate $5 + x$ when x is replaced by -8.

$5 + (-8) = -3$

5
Evaluate $3x - (-2)$ when $x = -3$.

$3(-3) + 2 = -7$

6
Evaluate $3 - (-3x)$ for $x = 4$.

$3 - (-3[4]) = 15$

7
Evaluate $4x - 3$ when $x = 4$.

13

8
Evaluate $a + 7$ when $a = -3$.

4

9
Evaluate $-2a - (-3)$ when $a = 7$.

-11

10
Evaluate $-2z - (-3)$ when $z = -5$.

13

11
Evaluate $-2 + (-4r)$ when $r = -3$.

10

12
Evaluate $6 + 2c$ when $c = -1$.

4

13
Evaluate $6x + 2$ for $x = -4$.

-22

14
Evaluate $5a + 3a$ for $a = 7$.

56

15
Evaluate $3(2x)$ for $x = -5$.

-30

16
Evaluate $2(y + 4)$ for $y = -3$.

2

17
Evaluate $-2x - (-5)$ for $x = -7$.

19

18
The open expression $x + 3y$ contains two letters. Replace x by -2 and y by 7 and evaluate the numerical expression.

$(-2) + 3(7) = 19$

19
Evaluate $2a + 3b$ when $a = -2$ and $b = 5$.

11

20
Evaluate $-3x + 5y + 2z$ when $x = -2$ and $y = 3$, and $z = -4$.

13

21
Evaluate $a - 3b$ when $a = -7$ and $b = -4$.

5

22
Evaluate $4x + 3c$ when $x = -1$ and $c = 4$.

8

23
Evaluate $2ab$ when $a = -2$ and $b = 5$.

-20

24
Evaluate $-17x + 3y$ when $x = 0$ and $y = -2$.

-6

25
Evaluate $-5x - (-4)$ when $x = -4$.

24

26
Evaluate $xy + 7$ when $x = -3$ and $y = 5$.

$(-3)(5) + 7 = -8$

27
Evaluate $9 - xy$ when $x = 4$ and $y = 0$.

9

28
Evaluate x^3 when $x = 2$.

$(2)(2)(2) = 8$

29
Evaluate x^3 when $x = 4$.

64

30
Evaluate y^2 when $y = -7$.

$(-7)(-7) = 49$

31
Evaluate a^4 when a = -2. 16

32
Evaluate z^2 when z = 1. 1

33
Evaluate x^4 when x = 3. 81

34
Evaluate x^3 when x = -3. -27

35
Evaluate $x^3 \cdot x$ when x = -2. $(-8) \cdot (-2) = 16$

36
Evaluate $a^2 \cdot a$ when a = 5. 125

37
Evaluate xx when x = -8. 64

38
Evaluate $x \cdot x^2$ when x = 3. 27

39
Evaluate $y^3 \cdot y^2$ when y = 2. 32

FEEDBACK UNIT 1

This quiz reviews the preceding unit. Answers are at the back of the book.

1. Evaluate $-7 + (-5x)$ when x = -2.
2. Evaluate $5 - r$ when r = -3.
3. Evaluate $2z - 3$ when z = -5.
4. Evaluate $4 + (3x - 6)$ when x = 8.
5. Evaluate $9 - (x + 3)$ when x = -13.
6. Evaluate x^3 when x = -5.
7. Evaluate x^4 when x = -3.
8. Evaluate $x^2 \cdot x^4$ when x = -2.

UNIT 2: SIMPLIFYING ADDITION EXPRESSIONS

The following mathematical terms are crucial to an understanding of this unit.

Mathematical statement	Open sentence
Commutative Law of Addition	Equivalent
Associative Law of Addition	Simplify
Like terms	

1

$2 + 3 \cdot 4 = 15 - 1$ is a true **mathematical statement**.
$5 + 8 \cdot -3 = -9 - 7$ is a false mathematical statement.
$4 \cdot 3 - 5 = (-3)(-3)$ is a ______ mathematical statement.

false

2

Is the statement $-3 \cdot 2 + 6 = 5 - 5$ true or false?

true

3

$x + 7 = 7 + x$ is an **open sentence**. It can be judged true or false when x is replaced by an integer. If x is replaced by 2, the statement $(2) + 7 = 7 + (2)$ is true. If $x = 3$, the statement $(3) + 7 = 7 + (3)$ is ______.

true

4

The open sentence $x - 3 = -3 + x$ becomes a ______ (true, false) statement when x is replaced by -5.

true

5

The open sentence $x + y = y + x$ becomes a ______ (true, false) statement when x is replaced by 2 and y is replaced by 4.

true

6

The open sentence $x + y = y + x$ becomes a ______ (true, false) statement when $x = -7$ and $y = -8$.

true

7

The open sentence $x + y = y + x$ becomes a ______ (true, false) statement for any integer replacements of x and y.

true

8

$x + y = y + x$ is a true statement for all integer replacements of x and y. This is the **Commutative Law of Addition**. The Commutative Law of Addition states $x + y =$ ______ for all integer replacements of x and y.

$y + x$

9

The Commutative Law of Addition, $x + y = y + x$, states that $x + y$ is **equivalent** to $y + x$ because the sum of addends is not affected by the order in which they are added. The Commutative Law of Addition states that $7 + 5 =$ _____.

$5 + 7$

10

By the Commutative Law of Addition, $a + 9$ is equivalent to ______.

$9 + a$

11

If $x = 5$ and $y = -4$, does the open sentence $(x + y) + 3 = x + (y + 3)$ become a true statement?

Yes

12

If $x = 9$, $y = -7$, and $z = 2$, does the open sentence $(x + y) + z = x + (y + z)$ become a true statement?

Yes

13

Does the open sentence $(x + y) + z = x + (y + z)$ become a true statement for any integer replacements for x, y, and z?

Yes

14

$(x + y) + z = x + (y + z)$ becomes a true statement for all integer replacements of x, y, and z. This is the **Associative Law of Addition**. The Associative Law of Addition states $(x + y) + z =$ ______.

$x + (y + z)$

15
The Associative Law of Addition states that $(x + y) + z$ is equivalent to $x + (y + z)$ because the sum of three addends is not affected by the grouping. The Associative Law of Addition states that $(3 + 5) + 7 =$ ______.

$3 + (5 + 7)$

16
By the Associative Law of Addition,
$(5 + 9) - 2 =$ ______.

$5 + (9 + [-2])$ or
$5 + (9 - 2)$

17
By the Associative Law of Addition,
$-12 + (-5 + 9) =$ ______.

$(-12 + [-5]) + 9$
or $(-12 - 5) + 9$

18
To **simplify** $(x + 5) - 2$ means to find an open expression that has the same evaluation for all integer replacements of x. Will the open expression $x + 3$ have the same evaluation as $(x + 5) - 2$ for all replacements of x?

Yes

19
To simplify $x - 8 + 6$ just add -8 and 6.

$$x - 8 + 6$$
$$x + (-8 + 6)$$
$$x + (-2)$$
$$x - 2$$

Simplify. $x - 8 - 7$

$x - 15$

20
To simplify $(4 + x) - 3$, both the Commutative and Associative Laws of Addition make the following steps possible.

$$(4 + x) - 3$$
$$(x + 4) - 3$$
$$x + (4 - 3)$$
$$x + 1$$

Simplify. $-7 + x + 2$

$x - 5$

21
Simplify. $8 + (-2 + x)$

$6 + x$

22
Simplify. $(8 + y) - 5$ — $y + 3$

23
The Commutative and Associative Laws of Addition allow any addition expression to be reordered and/or regrouped. To simplify $-5 + x + 7$, the following steps are used:

$$-5 + x + 7$$
$$x + (-5 + 7)$$
$$x + 2$$

Simplify. $-3 + x + 5$ — $x + 2$

24
To simplify $3x - 8 + 6$, add -8 and 6.

$$3x - 8 + 6$$
$$3x + (-8 + 6)$$
$$3x + (-2)$$

Is $3x + (-2)$ equal to $3x - 2$? — Yes

25
Simplify. $2x - 5 + 3$ — $2x - 2$

26
Simplify $3x - 8 + 10$ by adding the **like terms** -8 and 10. — $3x + 2$

27
Simplify. $8x - 3 + 7$ — $8x + 4$

28
Simplify. $3 - 5x - 7$ — $-4 - 5x$ or $-5x - 4$

29
Simplify $9 + 4x - 9$ by adding the like terms 9 and -9. — $4x$

30
Simplify. $3x - 4 + 8$ — $3x + 4$

31
Simplify. $7 + 5 - 3x$ — $12 - 3x$ or $-3x + 12$

32
Simplify. $4 + x + 9$ — $x + 13$

33
Simplify. $8x - 7 + 12$

$8x + 5$

34
Simplify. $7x + 8 - 3$

$7x + 5$

35
Simplify. $5x + 3 - 9$

$5x - 6$

36
Simplify. $2x - 7 + 9$

$2x + 2$

37
Simplify. $-3x + 4 - 9$

$-3x - 5$

38
Simplify. $3x - 4 + 4$

$3x$

FEEDBACK UNIT 2

This quiz reviews the preceding unit. Answers are at the back of the book.

1. Use the Commutative Law of Addition to write an equivalent expression for $8 + 3w$.
2. Use the Commutative Law of Addition to write an equivalent expression for $13x + 4y$.
3. Use the Associative Law of Addition to write an equivalent expression for $6x + (-8x + 3)$.
4. Use the Associative Law of Addition to write an equivalent expression for $(9x - 13) + 7$.
5. Find a simpler, equivalent expression for $5 - 9x - 12$.
6. Find a simpler, equivalent expression for $-6x - 14 - 5x$.
7. Simplify. $4 + 3x - 9$
8. Simplify. $-6 - 4x - 1$

UNIT 3: SIMPLIFYING MULTIPLICATION EXPRESSIONS

The following mathematical terms are crucial to an understanding of this unit.

Commutative Law of Multiplication   Associative Law of Multiplication

1
For the open sentence $x \cdot 3 = 3x$, if x is replaced by 7 will the statement be true?

Yes

2
For the open sentence $xy = yx$, if $x = -9$ and $y = 5$, the statement is true. If $x = 4$ and $y = -7$, is the statement obtained from $xy = yx$ true?

Yes

3
Does the open sentence $xy = yx$ become a true statement for all integer replacements of x and y?

Yes

4
$xy = yx$ is true for all integer replacements of x and y. This is the **Commutative Law of Multiplication**. The Commutative Law of Multiplication states that xy is equivalent to ______.

yx

5
By the Commutative Law of Multiplication, $x \cdot 3 = 3x$ and $x \cdot 4 =$ ______.

4x

6
By the Commutative Law of Multiplication, $z \cdot -5 = -5z$ and $y \cdot -7 =$ ______.

-7y

7
By the Commutative Law of Multiplication, $(3x)2 = 2(3x)$ and $(7x)5 =$ _____.

5(7x)

8
By the Commutative Law of Multiplication, $(-7y)6 = 6(-7y)$ and $(4z)8 =$ ______.

8(4z)

9

The Commutative Law of Multiplication, xy = yx, states that the product of two integers is not affected by the order in which they are multiplied. By the Commutative Law of Multiplication, -7 • 5 = _____.

5 • -7

10

By the Commutative Law of Multiplication, 3(-5) = ______.

(-5)3 or -5 • 3

11

If x = 4, does the open sentence -3(2x) = (-3 • 2)x become a true statement?

Yes

12

If x = 7 and y = -3, does the open sentence (4x)y = 4(xy) become a true statement?

Yes

13

If x = 4, y = -2, and z = 3, does the open sentence (xy)z = x(yz) become a true statement?

Yes

14

Does (xy)z = x(yz) become a true statement for all integer replacements of x, y, and z?

Yes

15

(xy)z = x(yz) is a true statement for all integer replacements of x, y, and z. This is the **Associative Law of Multiplication** which states that (xy)z is equivalent to ______.

x(yz)

16

By the Associative Law of Multiplication, 5(-2 • 6) = ______.

(5 • -2)6

17

The Associative Law of Multiplication, (xy)z = x(yz), states that the product of three integers is not affected by the grouping. The Associative Law of Multiplication states that (-3 • 5)4 = ______.

-3(5 • 4)

18

To simplify $5(-2x)$, the following steps are used.

$$5(-2x)$$
$$(5 \bullet -2)x$$
$$-10x$$

Simplify. $-3(4x)$ — $-12x$

19

Simplify. $8(3x)$ — $24x$

20

Simplify. $-3(6a)$ — $-18a$

21

Simplify. $-2(-3b)$ — $6b$

22

Simplify. $5(-3r)$ — $-15r$

23

Simplify. $-9(-7x)$ — $63x$

24

To simplify $(-4x)3$, both the order and the grouping of the factors must be altered.

$$(-4x)3$$
$$3(-4x)$$
$$(3 \bullet -4)x$$
$$-12x$$

Simplify. $(9x)6$ — $54x$

25

Simplify. $(5x) \bullet -4$ — $-20x$

26

Simplify. $(-8x) \bullet -2$ — $16x$

27

Simplify. $(-3x)7$ — $-21x$

28

Simplify. $5(-8x)$ — $-40x$

FEEDBACK UNIT 3

This quiz reviews the preceding unit. Answers are at the back of the book.

1. Use the Commutative Law of Multiplication to write an equivalent expression for (-7x) • 5.
2. Use the Commutative Law of Multiplication to write an equivalent expression for x • -8.
3. Use the Associative Law of Multiplication to write an equivalent expression for -6(5x).
4. Use the Associative Law of Multiplication to write an equivalent expression for (-4x) • x.
5. Find a simpler, equivalent expression for 5 • (7x).
6. Find a simpler, equivalent expression for -3 • (6x).
7. Simplify. -9 • (-3x)
8. Simplify. (-2x) • 7

UNIT 4: SIMPLIFYING OPEN EXPRESSIONS

The following mathematical terms are crucial to an understanding of this unit.

Multiplication Law of One
Distributive Law of Multiplication over Addition
Multiplication Law of Negative One

1

1 • 5 = 5, 1 • -7 = -7, and 1 • -12 = ______.

-12

2
The fact that any integer times 1 produces that same integer is so obvious that its usefulness is often overlooked. However, it often helps to introduce 1 as a factor. Is $9 + x$ equivalent to $9 + 1x$?

Yes

3
Do the open sentences $1x = x$ and $1 \cdot x = x$ become true statements for any integer replacement for x?

Yes

4
$1x = x$ is a true statement for any integer replacement of x. This is the **Multiplication Law of One** which states that 1x is equivalent to ______.

x

5
Is $4x + (3y - z)$ equivalent to $4x + 1 \cdot (3y - z)$?

Yes

6
By the Multiplication Law of One, $9 + (6x - 5)$ is equivalent to $9 + 1 \cdot (6x - 5)$. Is $1 + (2x - 7)$ equivalent to $(2x - 7)$?

No, it is equivalent to $1 + 1(2x - 7)$

7
By the Multiplication Law of One, $5 + x^3$ is equivalent to $5 + 1x^3$. Similarly, $8 + y^2$ is equivalent to ______.

$8 + 1y^2$

8
$9x + x$ is equivalent to $9x + 1x$. Similarly, $y - 6y$ is equivalent to ______.

$1y - 6y$

9
If -1 is multiplied by any integer it produces the opposite of the integer. $-1 \cdot 7 = -7$ and $-1 \cdot -5 = 5$. Is $-x + 14$ equivalent to $-1 \cdot x + 14$?

Yes, $-x = -1x$

10
$-1x = -x$ is a true statement for any integer replacement for x. $-z$ is equivalent to ______.

$-1z$

11
The Multiplication Law of Negative One states:
$-x$ is equivalent to $-1x$
Is $13 - k$ equivalent to $13 + (-k)$ or $13 + (-1k)$?

Yes

12
By the Multiplication Law of Negative One,
$-(3x - 5)$ is equivalent to $-1 \bullet (3x - 5)$
$-(2x + 9)$ is equivalent to ______.

$-1 \bullet (2x + 9)$

13
By the Multiplication Law of Negative One,
$-x^2$ is equivalent to $-1 \bullet x^2$ or $-1x^2$
Similarly, $-y^3$ is equivalent to ______.

$-1 \bullet y^3$

14
$6y - (6x + 5)$ is equivalent to $6y - 1(6x + 5)$.
Similarly, $10 - (-2x + 3)$ is equivalent to________.

$10 - 1(-2x + 3)$

15
The evaluation of $5(-7 + 3)$ is $5(-4)$ or -20.
What is the evaluation of $5 \bullet -7 + 5 \bullet 3$?

$-35 + 15 = -20$

16
The evaluation of $-6(4 - 9)$ is $-6(-5)$ or 30.
What is the evaluation of $-6 \bullet 4 + -6 \bullet -9$?

$-24 + 54 = 30$

17
Is the evaluation of $3(7 - 4)$ the same as the evaluation of $3 \bullet 7 + 3 \bullet -4$?

Yes, both are 9

18
Is the evaluation of $-4(1 - 6)$ the same as the evaluation of $-4(1) + [-4(-6)]$?

Yes, both are 20

19
The expression $a(b + c)$ is equivalent to $ab + ac$.
Find an equivalent expression for $k(m + r)$.

$km + kr$

20
The expression $x(y + z)$ is equivalent to $xy + xz$.
Find an equivalent expression for $j(t + s)$.

$jt + js$

21
Since $a(b + c)$ is equivalent to $ab + ac$,
$4(3x + 7)$ is equivalent to $4 \bullet 3x + 4 \bullet 7$
$5(x + 3)$ is equivalent to ____________

$5 \bullet x + 5 \bullet 3$ or $5x + 15$

22
Since ab + ac is equivalent to a(b + c),
9x + 5x is equivalent to (9 + 5)x
11x + 4x is equivalent to ___________

(11 + 4)x

23
The open sentence xy + xz = x(y + z) is a true statement for all integer replacements of x, y, and z. This is the **Distributive Law of Multiplication over Addition** which states that 5[3 + (-7)] will have the same evaluation as ______________.

5 • 3 + 5 • -7

24
Does 5(4 – 2) have the same evaluation as 5(4) – 5(2)?

Yes, 10

25
Which of the numerical expressions below has the same evaluation as 6(8 – 11)?
6(8) – 6(11) or 6(8) + 6(11)

6(8) – 6(11)

26
Which of the numerical expressions below has the same evaluation as -4(6 – 8)?
-4(6) – [-4(-8)] or -4(6) + 4(8)

-4(6) + 4(8)

27
Which of the numerical expressions below has the same evaluation as -7(-2 + 6)?
-7(-2) – [-7(6)] or -7(-2) + [-7(6)]

-7(-2) + [-7(6)]

28
To simplify 4x + 9x, the Distributive Law of Multiplication over Addition allows the addition of the **like terms**, 4x and 9x.

4x + 9x
(4 + 9)x
13x

Simplify. 8x + 3x

11x

29
To simplify the open expression 9x – 4x, add the like terms.

9x – 4x
(9 – 4)x
5x

Simplify. 12x – 8x

4x

30
Simplify. $7x - 9x$

$-2x$

31
Simplify. $5x - x$

$5x - 1x = 4x$

32
Simplify. $-3x - 6x$

$-9x$

33
Simplify. $x - 3x$

$1x - 3x = -2x$

34
Simplify. $x - x$

0

35
The addition expression $3x - 7 + 8x$ is simplified by adding the like terms $3x$ and $8x$.

$$3x - 7 + 8x$$
$$(3x + 8x) - 7$$
$$11x - 7$$

Simplify. $5x - 9 + 3x$

$8x - 9$

36
$6x + 7 - 2x$ is simplified by adding the like terms $6x$ and $-2x$.

$$6x + 7 - 2x$$
$$(6x - 2x) + 7$$
$$4x + 7$$

Simplify. $8x + 3 - 6x$

$2x + 3$

37
$4x - 6 - 9x$ is simplified as shown:

$$4x - 6 - 9x$$
$$(4x - 9x) - 6$$
$$-5x - 6$$

Simplify. $2x - 7 - 5x$

$-3x - 7$

38
Simplify. $x - 7 - 3x$

$-2x - 7$

39
Simplify. $12x - 3 + 2x$

$14x - 3$

40
Simplify. $7x + 3 - x$

$6x + 3$

41

To simplify $6x - 7 + 3$, the like terms -7 and +3 are added.

$$6x - 7 + 3$$
$$6x - 4$$

Simplify. $4x - 9 + 7$

$4x - 2$

42

To simplify $4x - 5 - 1$, add the like terms -5 and -1.

$$4x - 5 - 1$$
$$4x - 6$$

Simplify. $7x - 6 - 4$

$7x - 10$

43

To simplify $6x - 3 - 2x + 7$, pairs of like terms are added.

$$6x - 3 - 2x + 7$$
$$(6x - 2x) + (-3 + 7)$$
$$4x + 4$$

Simplify $4x - 8 - 2x + 3$ by adding the like terms.

$2x - 5$

44

To simplify $5x - 7 + x + 5$,

$$5x - 7 + x + 5$$
$$5x + x - 7 + 5$$
$$6x - 2$$

Simplify. $6x - 9 + x + 3$

$7x - 6$

45

Simplify. $4x + 5 - 7x + 2$

$-3x + 7$

46

Simplify. $5x - 6 - x - 9$

$4x - 15$

47

Simplify. $2x + 3 - 5 - 9x$

$-7x - 2$

48

Simplify. $2 - 3x - 5x + 8$

$-8x + 10$

49

Simplify. $7x - 9x - 5 - x$

$-3x - 5$

50

Simplify. $8 - 3x - 7 + 3x$

1

51
Simplify. $-6x - 5 + 9x - 1$

$3x - 6$

52
Simplify. $4x + 8 - x - 8$

$3x$

53
Simplify. $x - 9 - 2x - 3$

$-x - 12$

54
Simplify. $4x + 7 - 8x + 8x$

$4x + 7$

55
Simplify. $-8 + 2x - 2x + 8$

0

FEEDBACK UNIT 4

This quiz reviews the preceding unit. Answers are at the back of the book.

1. Use the Multiplication Laws of One and Negative One to write equivalent expressions for the following.
   a. x
   b. $-y$
   c. $6x - (5x - 3)$
   d. $-3 + (4x + 7)$
   e. $-x^4$
2. Simplify the following expressions.
   a. $6x - 2x$
   b. $-9x - x$
   c. $4x - 7 - 9x$
   d. $2x - 3 - 8x - 5$
   e. $7x + 5 + x - 13$

UNIT 5: REMOVING PARENTHESES TO SIMPLIFY OPEN EXPRESSIONS

The following mathematical term is crucial to an understanding of this unit.

Remove parentheses

1

To **remove parentheses** from $3(2x - 7)$, both $2x$ and -7 are multiplied by 3.

$$3(2x - 7)$$
$$3 \cdot 2x - 3 \cdot 7$$
$$6x - 21$$

Remove the parentheses. $5(4x - 9)$

$20x - 45$

2

The parentheses of $-6(2x + 5)$ are removed by multiplying both $2x$ and 5 by -6.

$$-6(2x + 5)$$
$$-6 \cdot 2x + (-6 \cdot 5)$$
$$-12x + (-30)$$

Is $-12x + (-30)$ equivalent to $-12x - 30$?

Yes

3

Remove the parentheses. $-7(2x + 5)$

$-14x - 35$

4

The parentheses of $-2(6x - 5)$ are removed by multiplying both $6x$ and -5 by -2.

$$-2(6x - 5)$$
$$-12x + 10$$

Remove the parentheses. $-8(2x - 3)$

$-16x + 24$

5

Remove the parentheses. $5(7x - 3)$

$35x - 15$

6

Remove the parentheses. $-3(8x - 1)$

$-24x + 3$

7
Remove the parentheses. $6(x - 1)$

$6x - 6$

8
Remove the parentheses. $-2(x + 7)$

$-2x - 14$

9
Remove the parentheses. $-8(5x - 2)$

$-40x + 16$

10
To remove the parentheses from $(2x - 3)$, the fact that $(2x - 3)$ is equivalent to $1 \bullet (2x - 3)$ is used.

$$(2x - 3)$$
$$1 \bullet (2x - 3)$$
$$2x - 3$$

Remove the parentheses. $(7x - 5)$

$7x - 5$

11
Remove the parentheses. $(8x + 3)$

$8x + 3$

12
To remove the parentheses from $-(3x - 5)$, the fact that $-(3x - 5)$ is equivalent to $-1 \bullet (3x - 5)$ is used.

$$-(3x - 5)$$
$$-1 \bullet (3x - 5)$$
$$-3x + 5$$

Remove the parentheses. $-(6x - 1)$

$-6x + 1$

13
Remove the parentheses. $-(2x + 9)$

$-2x - 9$

14
Remove the parentheses. $-7(x - 6)$

$-7x + 42$

15
To begin the simplification of $8 - 3(x - 4)$, first remove the parentheses.

$$8 - 3(x - 4)$$
$$8 - 3x + 12$$

Begin the simplification of $7 - 2(x + 8)$ by first removing the parentheses.

$7 - 2x - 16$

16

To simplify $7x + 4(2x - 5)$, first remove the parentheses; then add like terms.

$$7x + 4(2x - 5)$$
$$7x + 8x - 20$$
$$15x - 20$$

Simplify $2x + 5(3x - 7)$ by first removing the parentheses.

$2x + 15x - 35 = 17x - 35$

17

Simplify $8x - 7(x - 2)$ by first removing the parentheses.

$8x - 7x + 14 = x + 14$

18

Simplify. $-3 + 2(x + 5)$

$-3 + 2x + 10 = 2x + 7$

19

To simplify $5 - (3x - 6)$, first remove the parentheses using -1 as the multiplier.

$$5 - (3x - 6)$$
$$5 - 1(3x - 6)$$
$$5 - 3x + 6$$
$$-3x + 11$$

Simplify. $6x - (2x - 7)$

$6x - 2x + 7 = 4x + 7$

20

Simplify. $8 - (3x + 2)$

$8 - 3x - 2 = -3x + 6$

21

Remove the parentheses from $2 + (3x - 7)$ by using 1 as the multiplier.

$$2 + (3x - 7)$$
$$2 + 1(3x - 7)$$
$$2 + 3x - 7$$
$$3x - 5$$

Simplify $4 + (2x - 5)$ by first removing the parentheses.

$4 + 2x - 5 = 2x - 1$

22

The first step in simplifying $5 - 2(3x + 4)$ is to remove the parentheses.

$$5 - 2(3x + 4)$$
$$5 - 6x - 8$$
$$-6x - 3$$

Simplify. $7 - 3(2x - 4)$

$7 - 6x + 12 = -6x + 19$

23

To simplify $2(3x - 2) - 4(5x + 2)$, first remove both pairs of parentheses.

$$2(3x - 2) - 4(5x + 2)$$
$$6x - 4 - 20x - 8$$
$$-14x - 12$$

Simplify. $5(2x + 3) - 3(x + 5)$

$10x + 15 - 3x - 15 = 7x$

24

Simplify. $2(x - 7) + 3(5x - 3)$

$17x - 23$

25

Simplify. $2(4x - 3) - 2(x - 1)$

$6x - 4$

26

Simplify. $3(2x + 1) + 5(x - 9)$

$11x - 42$

27

To simplify $(5x - 3) - (2x + 2)$, first remove both pairs of parentheses.

$$(5x - 3) - (2x + 2)$$
$$1(5x - 3) - 1(2x + 2)$$
$$5x - 3 - 2x - 2$$
$$3x - 5$$

Simplify. $(2x + 3) - (x + 5)$

$x - 2$

28

Simplify. $(5 - 4x) - (x + 3)$

$-5x + 2$

29

Simplify. $-(3x + 7) - (x + 3)$

$-4x - 10$

30

Simplify. $(3x - 4) - (2x + 5)$

$x - 9$

31

Simplify. $2x - (5 - 3x)$

$5x - 5$

32

Simplify. $(4x - 3) - 2x$

$2x - 3$

33

Simplify. $3x + 5(x - 9)$

$8x - 45$

FEEDBACK UNIT 5

This quiz reviews the preceding unit. Answers are at the back of the book.

1. Simplify. $7(3x - 5)$
2. Simplify. $-4(x - 3)$
3. Simplify. $5 + 2(3x - 7)$
4. Simplify. $8 - (3 + 5x)$
5. Simplify. $4x + 3(2x - 1)$
6. Simplify. $(2x - 5) - (x + 7)$
7. Simplify. $3(2x - 1) + 2(x + 3)$

UNIT 6: SIMPLIFYING POWER EXPRESSIONS

The following mathematical terms are crucial to an understanding of this unit.

Squared	Second power	Exponent
Third power	Base	Fourth power
Power expression	Coefficient	Simplifying power multiplications

1

x^2 is read "x **squared**" or "x to the **second power**." The **exponent** is 2 in x^2 which means $x \cdot x$. Evaluate x^2 when $x = 5$.

$5^2 = 5 \cdot 5 = 25$

2

x^3 is read "x to the **third power**." The **base** is x in x^3 which means $x \cdot x \cdot x$. Evaluate x^3 when $x = 2$.

$2 \cdot 2 \cdot 2 = 8$

3

When $x = 2$, do the open expressions $x \cdot x^2$ and x^3 have the same evaluation?

Yes, both are 8

4

a^4 is read "a to the **fourth power**." Evaluate a^4 when $a = -2$.

16

5

x^3 is a **power expression** with base x and exponent 3. Write the power expression with base y and exponent 5.

y^5

6

Write the power expression with base k and exponent 4.

k^4

7

Write the power expression with base m and exponent w.

m^w

8

Write the power expression with base 3 and exponent 6.

3^6

9

The power expression with base x and exponent 1 can be written as x^1, but usually is written without the exponent. Write the power expression with base 7 and exponent 4.

7^4

10

In the power expression x^3, the exponent indicates that the base is used 3 times as a factor. Is x^3 equivalent to xxx?

Yes

11

In the power expression z^8, the exponent indicates that the base, ____, is used _____ times as a factor.

z, 8

12

In the power expression y^6, the exponent indicates that the base, ____, is used _____ times as a factor.

y, 6

13

Write the power expression in which k is a factor 4 times.

k^4

14

In the power expression x^7, how many times is x used as a factor?

7

15
In the open expression x^5, the exponent 5 shows that x is to be used as a factor 5 times. How many factors of x are in x^3?

3

16
x^4 means 4 factors of x, or xxxx.
x^2 means 2 factors of x, or xx.
x^3 means 3 factors of x, or ______.

xxx

17
y^3 means 3 factors of y, or yyy.
y^5 means 5 factors of y, or ______.

yyyyy

18
a^5 means 5 factors of a, or aaaaa.
a^4 means 4 factors of a, or ______.

aaaa

19
$7x^3$ means 7 • xxx or 7xxx. The 7 is a **coefficient**. $5x^5$ means __________.

5 • xxxxx or 5xxxxx

20
$6a^3$ means 6aaa. The coefficient, 6, is used ____ time as a factor while a is used three times as a factor.

one

21
x^2 means 1xx, but the +1 is often not shown. What is the coefficient of x^4?

1 or +1

22
$-x^4$ means -1 • xxxx or -1xxxx.
The coefficient is _______.

-1

23
$-3x^4$ means ________.

-3xxxx

24
$-x^5$ means _______.

-1xxxxx

25
x^3 means _______.

1xxx or xxx

26
$x^4 \bullet x^3$ means xxxx • xxx. How many factors of x are in the open expression xxxx • xxx?

7

27
The multiplication of x^2 and x^3 is shown by $x^2 \cdot x^3$ and means $xx \cdot xxx$. How many factors of x are in the open expression $x^2 \cdot x^3$?

5

28
$x^3 \cdot x$ means $xxx \cdot x$. How many factors of x are in the open expression $x^3 \cdot x$?

4

29
$a^5 \cdot a^3$ means $aaaaa \cdot aaa$. How many factors of a are in the open expression $a^5 \cdot a^3$?

8

30
How many factors of the base a are in the open expression $a^3 \cdot a^2$?

5

31
How many factors of the base x are in the open expression $x^5 \cdot x^4$?

9

32
To multiply two power expressions like x^7 and x^3, count the total number of factors of the base. How many factors of the base x are in $x^7 \cdot x^3$?

10

33
To multiply two power expressions with the same base, count the total number of factors of the base. How many factors of the base x are in $x \cdot x$?

2

34
To simplify the multiplication of two power expressions, use the meaning of the exponents to count how many factors of the base are involved. How many factors of r are in $r^2 \cdot r^4$?

6

35
How many factors of the base x are in the open expression $x \cdot x^8$?

9

36
The multiplication of x^6 and x^3 can be simplified as follows.

$$x^6 \cdot x^3 = xxxxxx \cdot xxx = x^9$$

Does $xxxxxx \cdot xxx$ contain 9 factors of x?

Yes

37
The multiplication of x^2 and x^4 can be simplified as follows.

$x^2 \bullet x^4 = xx \bullet xxxx = x^6$

Does $xx \bullet xxxx$ contain 6 factors of x?

Yes

38
The multiplication of a^7 and a^4 can be simplified as follows.

$a^7 \bullet a^4 = aaaaaaa \bullet aaaa = a^{11}$

Does $aaaaaaa \bullet aaaa$ contain 11 factors of a?

Yes

39
Should the multiplication of $x^2 \bullet x^7$ be simplified to x^{14}?

No, it is x^9

40
Should the multiplication of $x^4 \bullet x^3$ be simplified to x^{12}?

No, it is x^7

When multiplying power expressions with the same base, the exponents are added.

$x^2 \bullet x^7 = x^{2+7} = x^9$ $y^5 \bullet y^8 = y^{5+8} = y^{13}$ $z \bullet z^6 = z^1 \bullet z^6 = z^7$

41
To **simplify the multiplication of power expressions** with the same base, the following steps are used:

$x^3 \bullet x^4 = x^{3+4} = x^7$

Simplify. $x^2 \bullet x^3$

$x^{2+3} = x^5$

42
To simplify $x^8 \bullet x^3$, the following steps are used:

$x^8 \bullet x^3 = x^{8+3} = x^{11}$

Simplify. $x^4 \bullet x^5$

x^9

43
Complete the simplification: $x \bullet x^4 =$ _______

x^5

44
Simplify. $x \bullet x^6$

x^7

45
Simplify. $x \cdot x^3$

x^4

46
Simplify. $x^3 \cdot x^6$

x^9

47
Simplify. $x^4 \cdot x$

x^5

48
Simplify. $x^7 \cdot x^4$

x^{11}

49
Simplify. $x^8 \cdot x^2$

x^{10}

50
Simplify. $r^3 \cdot r^5$

r^8

51
Simplify. $x^{13} \cdot x^3$

x^{16}

52
Simplify. $a^5 \cdot a^7$

a^{12}

53
Simplify. $x \cdot x$

x^2

54
Simplify. $x^2 \cdot x$

x^3

55
Simplify. $x^2 \cdot x^2$

x^4

FEEDBACK UNIT 6

This quiz reviews the preceding unit. Answers are at the back of the book.

1. Use the meaning of an exponent to write an equivalent expression for $x \cdot x \cdot x \cdot x \cdot x$.
2. Use the meaning of an exponent to write an equivalent expression for y^4.
3. Simplify. $x^5 \cdot x^4$
4. Simplify. $y^{13} \cdot y^9$
5. Simplify. $z^8 \cdot z^4$
6. Simplify. $x^4 \cdot x^3$
7. Simplify. $x^8 \cdot x$
8. Simplify. $x \cdot x$

UNIT 7: APPLICATIONS

In this Applications Section, the format of the text has been altered. Answers for the problems appear beneath them rather than in the right-hand column. Your studying emphasis should be on learning the best procedures to follow with word problems.

1

There is a need to learn how to translate words and phrases into open expressions.

a number plus 3	translates as	N + 3
7 more than a number	translates as	N + 7
a number increased by 8	translates as	N + 8

Write a translation for: 11 added to a number

Answer: N + 11 is correct. Another correct answer is 11 + N. With addition expressions the order will not affect the correctness.

2

In translating subtraction phrases into open expressions, be careful about the correct order of the terms.

8 subtracted from a number	translates as	N – 8
a number minus 13	translates as	N – 13
12 decreased by a number	translates as	12 – N

Write a translation for: 5 less than a number

Answer: N – 5 is correct. 5 – N is incorrect.

3

The open expression 3N could be the translation of:

3 multiplied by a number
the product of 3 and a number
a number increased threefold

The open expression 2N could be a translation of which of the following?

a. twice a number
b. the quotient of 2 and N
c. a number doubled

Answer: (a) and (c) are good translations, but (b) indicates a division.

4

To write an open expression for: 12 more than a number

a. The phrase "more than" is a clue that the operation is addition
b. The translation is direct.

12	more than	a number
12	+	N

Either 12 + N or N + 12 are correct. Write an open expression for: a number increased by 17

Answer:

a. The clue phrase is "increased by" and indicates addition.
b.

a number	increased by	17
N	+	17

Either N + 17 or 17 + N are correct open expressions.

5

To write an open expression for: 9 less than a number

a. The phrase "less than" is a clue that the operation is subtraction

b. When dealing with subtraction, the order of the terms may need to be reversed. For this open expression, that is the case.

9	less than	a number
N	–	9

N – 9 is the correct result. Write an open expression for: 13 subtracted from a number

Answer: a. The clue phrase is "subtracted from" and indicates subtraction.

b.

13	subtracted from	a number
N	–	13

N – 13 is the correct open expression. 13 – N is incorrect.

6

To write an open expression for: 6 multiplied by a number

a. Notice that the operation is multiplication.

b. The translation is direct.

6	multiplied by	a number
6	•	N

6 • N or 6N is the correct result. Write an open expression for: 4 times a number

Answer: a. The clue word is "times" and indicates multiplication.

b.

4	times	a number
4	•	N

4 • N or 4N is a correct open expression.

FEEDBACK UNIT 7 FOR APPLICATIONS

1. Write an open expression for: a number increased by 14.

2. Write an open expression for: 15 less than a number.

3. Write an open expression for: 7 multiplied by a number.

4. Write an open expression for: 19 more than a number .

5. Write an open expression for: a number decreased by 11.

SUMMARY FOR THE ALGEBRA OF THE INTEGERS

The following mathematical terms are crucial to an understanding of this chapter.

Open expression
Evaluation
Open sentence
Equivalent
Commutative Law of Multiplication
Associative Law of Multiplication
Multiplication Law of Negative One
Like terms
Remove parentheses
Squared
Base
Fourth power
Simplifying power multiplications
Numerical expression
Mathematical statement
Commutative Law of Addition
Associative Law of Addition
Simplify
Multiplication Law of One
Distributive Law of Multiplication over Addition
Exponent
Second power
Third power
Power expression
Coefficient

In Chapter 5 the simplification of expressions involving the integers has been studied. The set of integers has all the properties of the set of counting numbers, but it includes some new properties.

As in the set of counting numbers, the properties of the integers include:

1. The Commutative and Associative Laws of Addition and Multiplication.
2. The Multiplication Law of One.
3. The Distributive Law of Multiplication over Addition.

As new properties, the set of integers has:

1. Zero as a number.
   a. The sum of any integer and zero is that integer.
   b. The product of any integer and zero is zero.
2. Every integer has an opposite. The sum of any integer and its opposite is zero.
3. Any integer multiplied by -1 gives its opposite.

Using the foregoing properties of integers, the methods for simplifying algebraic expressions such as $3x - 8 - 2x$, $3(2x - 3)$, $-(5x + 1)$, and $x^2 \cdot x^5$ were studied.

CHAPTER 5 MASTERY TEST

The following questions test the objectives of Chapter 5. Answers are at the back of the book. The number in parentheses which follows each problem indicates the unit in which it can be learned.

Evaluate each of the following expressions.

1. $7 - 3x$, when $x = -4$ (1)

2. $xy - 3$, when $x = 2$ and $y = -4$ (1)

3. $x \cdot x^4$, when $x = -3$ (1)

Complete the following statements.

4. By the Commutative Law of Addition, $4x + 3 =$ ______. (2)

5. By the Associative Law of Addition, $9 + (7x + 5) =$ ______. (2)

Simplify each of the following expressions.

6. $8 - 3x + 9$ (2)

7. $6x - 3 - 6x$ (2)

8. $3 \cdot (-4x)$ (3)

9. $(-3x) \cdot 9$ (3)

10. $x \cdot x^7$ (6)

11. $x^4 \cdot x^3$ (6)

12. $x^9 \cdot x$ (6)

Complete the following statements.

13. By the Associative Law of Multiplication, $4(3 \cdot 5) =$ ______. (3)

14. By the Commutative Law of Multiplication, $-2x =$ ______. (3)

Simplify each of the following expressions.

15. $-8 + 4x - 2 - 9x$ (4)

16. $-5(3x - 7)$ (5)

17. $3x - (2x - 3)$ (5)

18. $2(4x - 3) - (7x + 4)$ (5)

19. $-(2x - 7) - (x + 13)$ (5)

20. $(3 - 5x) + 5(x - 3)$ (5)

21. Write an open expression for: a number diminished by 6. (7)

22. Write an open expression for: the sum of a number and 8. (7)

Chapter 6 Objectives

The following problems illustrate the objectives of this chapter. At this time you are not expected to know how to do these problems. However, if all of these problems are thoroughly understood, proceed directly to the Chapter 6 Mastery Test. The number in parentheses which follows each problem indicates the unit in which it can be learned.

Find solutions for the following equations.

1. $x + 8 = 5$ (1)

2. $x - 13 = 19$ (1)

3. $7y = -56$ (1)

4. $-4y = -32$ (1)

5. $7x + 13 = 55$ (2)

6. $6x - 7 = 3x + 14$ (2)

7. $3 - 8x = -3x - 12$ (2)

8. $3(2x - 5) + 4 = 7$ (3)

9. $-7x + 4 = 2x - 23$ (3)

10. $4(x - 9) + 5x = 18$ (3)

Find the truth set and show a check for each of the following equations.

11. $4x - 3 = 25$ (4)

12. $3x - 8 = 5x + 4$ (4)

13. $-x + 8 = 2x - 1$ (4)

14. $3(x - 5) = x - 7$ (4)

CHAPTER 6
SOLVING EQUATIONS WITH THE INTEGERS

UNIT 1: SOLVING SIMPLE EQUATIONS

The following mathematical terms are crucial to an understanding of this unit.

Equation Statement

1
In the **equation** $x + 4 = 7$, if x is replaced by 5 the resulting statement, $5 + 4 = 7$, is false. If x is replaced by 9 in $x + 4 = 7$, the statement $9 + 4 = 7$ is ______. (true or false)

false

2
In the equation $x + 2 = 9$, if x is replaced by 7 the **statement** $7 + 2 = 9$ is ______. (true or false)

true

3
In the equation $x + 7 = -3$, if x is replaced by -1 the statement $-1 + 7 = -3$ is false. In the equation $x + 7 = -3$, if x is replaced by -10 the statement $-10 + 7 = -3$ is ______. (true or false)

true

4
The equation $x + 3 = 7$ can be translated into a question:
What integer can be added to 3 and equal 7?
The answer to the question is ______.

4

5
The equation $x - 8 = 5$ can be translated into a question:
What integer can be added to -8 and equal 5?
The answer to the question is ______.

13

6
Find the integer that will make $x + 8 = 11$ a true statement.

3

7
Find the integer that will make $x + 7 = 5$ a true statement.

-2

8
Find the integer that will make $x - 1 = -5$ a true statement.

-4

9
Find the integer that will make $5 = x + 8$ a true statement.
[Note: $5 = x + 8$ can be written as $x + 8 = 5$.]

-3

10
Find the integer that will make $x + 9 = 17$ a true statement.

8

11
Find the integer that will make $x + 5 = 1$ a true statement.

-4

12
Find the integer that will make $x - 1 = -4$ a true statement.

-3

13
Find the integer that will make $-3 = x - 5$ a true statement.

2

14
Find the integer that will make $x - 7 = -4$ a true statement.

3

15
Find the integer that will make $x + 2 = -3$ a true statement.

-5

16
Find the integer that will make $5 = x - 3$ a true statement.

8

17
Find the integer that will make $x - 7 = -7$ a true statement.

0

18
Find the integer that will make $x + 3 = -5$ a true statement.

-8

19
Find the integer that will make $x + 2 = 13$ a true statement.

11

20
The equation $5x = 10$ translates into a multiplication question.

What integer can be multiplied by 5 and equal 10?

The answer to the question is ______.

2

21
The equation $3x = 18$ translates into a multiplication question.

What integer can be multiplied by 3 and equal 18?

The answer to the question is ______.

6

22
The equation $-4x = 20$ translates into a multiplication question.
What integer can be multiplied by -4 and equal 20?
The answer to the question is ______. -5

23
Find the integer that will make $-2x = 12$ a true statement. -6

24
Find the integer that will make $8x = 0$ a true statement. 0

25
Find the integer that will make $-2x = -6$ a true statement. 3

26
Find the integer that will make $-20 = 5x$ a true statement. -4
[Note: $-20 = 5x$ may be written as $5x = -20$.]

27
Find the integer that will make $7x = -35$ a true statement. -5

28
Find the integer that will make $-7x = 21$ a true statement. -3

29
Find the integer that will make $-3x = -12$ a true statement. 4

30
Find the integer that will make $7 = -7x$ a true statement. -1

31
Find the integer that will make $5x = -30$ a true statement. -6

32
Find the integer that will make $-15 = -5x$ a true statement. 3

33
To find the integer that will make $x + 9 - 5 = 11$ a true statement, first simplify the **left side** of the equation

$$x + 9 - 5 = 11$$
$$x + 4 = 11$$

Since $x + 9 - 5$ and $x + 4$ are equivalent expressions, find the integer that will make $x + 4 = 11$ a true statement.

7

34
Find the integer that will make $x + 8 - 6 = 7$ a true statement by first simplifying $x + 8 - 6$.

5

35
Find the integer that will make $x - 9 + 4 = 7$ a true statement by first simplifying $x - 9 + 4$.

12

36
To find the integer that will make $3x + 4 - 2x = 7$ a true statement, first simplify the left side of the equation

$$3x + 4 - 2x = 7$$
$$x + 4 = 7$$

Since $3x + 4 - 2x$ and $x + 4$ are equivalent expressions, find the integer that will make $x + 4 = 7$ a true statement.

3

37
Find the integer that will make $5x - 2 - 4x = 3$ a true statement by first simplifying $5x - 2 - 4x$.

5

38
To find the integer that will make $5x - (4x - 7) = -3$ a true statement, the following steps are used:

$$5x - (4x - 7) = -3$$
$$5x - 1(4x - 7) = -3$$
$$5x - 4x + 7 = -3$$
$$x + 7 = -3$$

Find the integer that will make $5x - (4x - 7) = -3$ a true statement.

-10

[Note: $5x - (4x - 7)$ is equivalent to $x + 7$.]

39
Find the integer that will make $12 = 9 - (9 - 3x)$ a true statement.

4

40
Find the integer that will make
$12 + (5x - 12) = 10$ a true statement.

2

41
Find the integer that will make
$5x + 3 - 4x = 7$ a true statement.

4

FEEDBACK UNIT 1

This quiz reviews the preceding unit. Answers are at the back of the book.

Find an integer that will make each equation a true statement.

1. $x + 8 = 23$
2. $x - 11 = 13$
3. $x + 6 = -15$
4. $x - 6 = -41$
5. $6x = 42$
6. $-7x = 35$
7. $-x = -9$
8. $9x - (8x - 7) = 17$
9. $12 - (12 - 3x) = -24$
10. $19 = (3 - x) - 3$

UNIT 2: GENERATING EQUIVALENT EQUATIONS

The following mathematical term is crucial to an understanding of this unit.

Equivalent equations

1
Two equations are equivalent, if they become true statements for the same integer replacements of the variable. Are $x + 3 = 5$ and $x + 3 + 4 = 5 + 4$ **equivalent equations**?

Yes, 2 makes both true.

2

If two equations become true for the same integer replacements of x, they are equivalent equations. Are $x + 13 = 7$ and $x + 13 - 9 = 7 - 9$ equivalent equations?

Yes, -6 makes both true.

3

Any integer may be added to **both sides of an equation** to generate a new, equivalent equation. Is $x + 4 = 9$ equivalent to $x + 4 + 12 = 9 + 12$?

Yes, 12 was added to both sides of the equation.

4

To solve $2x + 5 = 13$, generate an equivalent equation by adding -5 to both sides of the equation.

$$\begin{array}{rcr} 2x + 5 & = & 13 \\ \underline{\quad -5} & & \underline{-5} \\ 2x + 0 & = & 8 \\ 2x \quad & = & 8 \end{array}$$

$2x + 5 = 13$ and $2x = 8$ are equivalent equations and both become true statements when x = _____.

4

5

To solve $3x - 7 = 8$, first generate an equivalent equation by adding 7 to both sides.

$$\begin{array}{rcr} 3x - 7 & = & 8 \\ \underline{\quad +7} & & \underline{+7} \\ 3x + 0 & = & 15 \\ 3x \quad & = & 15 \end{array}$$

$3x - 7 = 8$ and $3x = 15$ are equivalent equations and both become true statements when x = _____.

5

6

To solve $3 + 2x = -9$, first generate an equivalent equation.

$$\begin{array}{rcr} 3 + 2x & = & -9 \\ \underline{-3 \quad\quad} & & \underline{-3} \\ 0 + 2x & = & -12 \\ 2x & = & -12 \end{array}$$

$3 + 2x = -9$ and $2x = -12$ are equivalent equations and both become true statements when x = _____.

-6

7

To solve $5x - 2 = 13$, the following steps are used.

$$\begin{array}{rcr} 5x - 2 & = & 13 \\ \underline{\quad +2} & & \underline{+2} \\ 5x + 0 & = & 15 \\ 5x \quad & = & 15 \end{array}$$

The integer that will make $5x - 2 = 13$ true is _____.

3

8

To solve $15 - 4x = 7$, the following steps are used.

$$\begin{array}{rcr} 15 - 4x & = & 7 \\ \underline{-15} & & \underline{-15} \\ 0 - 4x & = & -8 \\ -4x & = & -8 \end{array}$$

The integer that will make $15 - 4x = 7$ true is _____.

2

9

To solve $2x + 7 = -3$, the following steps are used.

$$\begin{array}{lcr} 2x + 7 & = & -3 \\ \underline{\quad\; -7} & & \underline{-7} \\ 2x + 0 & = & -10 \\ 2x & = & -10 \end{array}$$

The integer that will make $2x + 7 = -3$ true is _____.

-5

10

To solve $5 = 2x - 3$, the following steps are used.

$$\begin{array}{rcl} 5 & = & 2x - 3 \\ \underline{+3} & & \underline{\quad +3} \\ 8 & = & 2x + 0 \end{array}$$

The integer that will make $5 = 2x - 3$ true is _____.

4

11

Find the integer that will make $8 - 3x = 2$ true by generating a simpler, equivalent equation.

$-3x = -6$
$x = 2$

12

Find the integer that will make $5x - 2 = 18$ true by generating a simpler, equivalent equation.

$5x = 20$
$x = 4$

13

Find the integer that will make $8x + 3 = 35$ true by generating a simpler, equivalent equation.

$8x = 32$
$x = 4$

14

Find the integer that will make $5 - 7x = -16$ true by generating a simpler, equivalent equation.

$-7x = -21$
$x = 3$

15

Find the integer that will make $14 = 2x + 4$ true by generating a simpler, equivalent equation.

$10 = 2x$
$x = 5$

16
Find the integer that will make $9 - x = 4$ true by generating a simpler, equivalent equation.

$-1x = -5$
$x = 5$

17
Find the integer that will make $-4x + 13 = 1$ true by generating a simpler, equivalent equation.

$-4x = -12$
$x = 3$

In the set of integers every number has an opposite. A number can be eliminated from one side of an equation by adding its opposite. It is essential to add the number to both sides of the equation to obtain an equivalent equation.

18
Find the integer that will make $-5x + 3 = 13$ true.

-2

19
Find the integer that will make $-3x + 7 = 22$ true.

-5

20
Find the integer that will make $-17 = 3x - 8$ true.

-3

21
Find the integer that will make $14 = 6 - 2x$ true.

-4

22
Find the integer that will make $4 - 5x = -11$ true.

3

23
Find the integer that will make $9x - 3 = 15$ true.

2

24
Find the integer that will make $8x - 3 = 21$ true.

3

25
Find the integer that will make $3 - x = 1$ true.

2

26
Find the integer that will make $9 = 2x - 5$ true.

7

27
To solve $5x + 3 + 2x = 31$, first simplify $5x + 3 + 2x$. Then find the integer that will make $5x + 3 + 2x = 31$ true.

$7x + 3 = 31$
$x = 4$

28
Find the integer that will make $3x - 4x + 12 = 7$ true by first simplifying the left side of the equation.

$-1x = -5$
$x = 5$

29
Find the integer that will make $6x - 3 - 2x = 9$ true.

3

30
Find the integer that will make $8x + 5 - 3x = 15$ true.

2

31
Find the integer that will make $5 - 2x = 3$ true.

1

FEEDBACK UNIT 2

This quiz reviews the preceding unit. Answers are at the back of the book.

Find an integer that will make each equation true.

1. $x + 6 = -1$
2. $-5x = 20$
3. $4x + 3 = 31$
4. $5 - 2x = 13$
5. $-9 = 7 + 8x$
6. $8x - 7 = 17$
7. $6x + 17 = -13$
8. $2x + 5 + 4x = 35$
9. $6x - 7 - 4x = 11$
10. $5x + 4 - 3x = -14$
11. $14 - x = 3$
12. $5 - x = -3$

UNIT 3: FINDING SOLUTIONS

The following mathematical terms are crucial to an understanding of this unit.

Solution Solve

1

For $5x = 9 + 2x$, the variable x is on both sides of the equation. To solve, first add the opposite of $2x$ to both sides of the equation.

$$\begin{array}{rl} 5x &= 9 + 2x \\ \underline{-2x} & \underline{\quad -2x} \\ 3x &= 9 + 0 \\ 3x &= 9 \end{array}$$

Find a **solution** (integer that will make it true) for $5x = 9 + 2x$. 3

2

To **solve** (find a solution) $2x = 6 - x$, begin by adding x to both sides of the equation.

$$\begin{array}{rl} 2x &= 6 - x \\ \underline{+x} & \underline{\quad +x} \\ 3x &= 6 + 0 \\ 3x &= 6 \end{array}$$

Find a solution for $2x = 6 - x$. 2

3

Find a solution for $8x = 12 + 6x$ by first adding $-6x$ to both sides of the equation. 6

4

Find a solution for $5x = 16 - 3x$ by first adding $3x$ to both sides of the equation. 2

5

Find a solution for $2x - 6 = 4x$ by first adding $-2x$ to both sides of the equation. -3

6

Find a solution for $3x - 5 = 8x$. -1

7

To solve $5x + 6 = 3x - 2$, begin by adding $-3x$ to both sides of the equation.

$$\begin{array}{l} 5x + 6 = 3x - 2 \\ \underline{-3x \qquad -3x} \\ 2x + 6 = 0 - 2 \end{array}$$

Complete the solution for $5x + 6 = 3x - 2$ by finding a solution of $2x + 6 = -2$.

-4

8

To solve $5x - 3 = 2x + 9$, the process can begin by adding either 3 or $-2x$ to both sides of the equation. Add 3 to both sides of $5x - 3 = 2x + 9$. What equivalent equation is obtained?

$5x = 2x + 12$

9

To solve $5x - 3 = 2x + 9$, the process can begin by adding either 3 or $-2x$ to both sides of the equation. Add $-2x$ to both sides of $5x - 3 = 2x + 9$. What equivalent equation is obtained?

$3x - 3 = 9$

10

In frames 8 and 9, two different approaches were taken for solving $5x - 3 = 2x + 9$. Solve both $5x = 2x + 12$ and $3x - 3 = 9$. Do they have the same solution?

Yes, 4

11

Find a solution for $3x + 7 = x - 9$. Begin the process by adding either -7 or $-x$ to both sides of the equation.

-8

12

Find a solution for $4 - 5x = -3x + 6$. Begin the process by adding either -4 or $3x$ to both sides of the equation.

-1

13

Find a solution for $2x - 7 = 5x + 2$.

-3

14

Find a solution for $3x - 6 = 4x - 10$.

4

15

Find a solution for $4x - 7 = x + 11$.

6

16
Find a solution for $2x - 9 = -3x + 11$.
4

17
Find a solution for $2x + 3 = 9x - 4$.
1

18
Find a solution for $12x - 20 = 3x + 7$.
3

19
Find a solution for $5x + 7 = 2x + 7$.
0

20
Find a solution for $-5x + 3 = 9 - 3x$.
-3

21
Find a solution for $-2x - 7 = 5x + 14$.
-3

22
Find a solution for $3x + 8 = 14x + 8$.
0

23
Find a solution for $11 - 2x = x + 8$.
1

24
Find a solution for $7x - 3 = 10x + 9$.
-4

25
Find a solution for $5x - 3 = 2x + 3$.
2

26
To solve $8x + 7 - 2x = 3x + 19$, first simplify the left side of the equation. After this simplification, what new equation is obtained?
$6x + 7 = 3x + 19$

27
Find a solution for $9x + 2 - 7x = x + 8$.
[Note: First simplify the left side of the equation.]
6

28
Find a solution for $2x - 3 = 5x + 7 - 4x$.
[Note: First simplify the right side of the equation.]
10

29
Find a solution for $3x + 5 - x = 8 - x + 3$.
2

30
Find a solution for $7x - 4 - 3x = 5x + 9 + 7$. -20

31
Find a solution for $5x - 2 - 3x = x + 3$. 5

32
Find a solution for $5x - 7 + x = 2x - 3 + 8$. 3

33
Find a solution for $8x - 4 + 2x = 10 + 3x$. 2

34
Find a solution for $8 - 3x + 3 = 7x - 5 - 2x$. 2

35
Find a solution for $x + 9 - 2x = 7 - 3x$. -1

36
Find a solution for $5x - 2x + 7 = 8x + 4 - 6x$. -3

37
The first step in finding a solution for $2(3x - 4) + 5 = 2x + 5$ is to remove the parentheses.

$$2(3x - 4) + 5 = 2x + 5$$
$$6x - 8 + 5 = 2x + 5$$
$$6x - 3 = 2x + 5$$

Complete the solution for $2(3x - 4) + 5 = 2x + 5$. 2

38
$3(5 - 2x)$ is equivalent to $15 - 6x$.
Find a solution for $8x + 3(5 - 2x) = 3$. -6

39
Find a solution for $5 - 3(2x + 4) = 11$. -3

40
Find a solution for $9 + 5(x + 4) = 2x - 2(x + 3)$.
[Note: Simplify both sides of the equation by removing the parentheses and combining like terms.] -7

41
Find a solution for $7 - (2x + 3) = 2(3x - 1) - 2$. 1

42
Find a solution for $5(2x - 3) = 6x + 1$. 4

43
Find a solution for $2(3x - 1) \doteq 5(x + 2) - 7$. 5

44
Solve. $x + 47 = 52$
[Note: "**Solve**" means to find a solution.] 5

45
Solve. $x - 15 = 63$ 78

46
Solve. $-8x + 5 = -27$ 4

47
Solve. $7 - x = 54$ -47

48
Solve. $7x - 5 = -68$ -9

49
Solve. $19x + 7 = 7$ 0

50
Solve. $4x - 5 = 19$ 6

51
Solve. $-9 = 3 + 6x$ -2

52
Solve. $2x + 3 = x - 7$ -10

53
Solve. $5 - 3x = -7$ 4

54
Solve. $4x = x + 15$ 5

55
Solve. $7 = 3 - 2x$ -2

56
Solve. $19 = 12 + 7x$ 1

57
Solve. $9 - x = 5 - 2x$ -4

The first step in solving any equation is to simplify each side of the equation.

$$3 - (5x - 4) = -8$$
$$3 - 5x + 4 = -8$$
$$-5x + 7 = -8$$

58
Find a solution for $4 + 3(x + 1) = 31$. 8

59
Find a solution for $2(3x - 7) + 4 = 2$. 2

60
Find a solution for $5x + 4 = 4 - 2x$. 0

61
Solve. $2(x - 3) = 12$ 9

62
Solve. $5x + 3(x + 1) = 19$ 2

63
Solve. $x - 9 = 5x + 11$ -5

64
Solve. $2x + 3 - x = 2x - 8$
[Note: Simplify the left side of the equation first.] 11

65
Solve. $8 - (2x - 3) = 17$
[Note: $-(2x - 3)$ is equivalent to $-1(2x - 3)$.] -3

66
Solve. $2x - (x - 4) = -9$ -13

67
Solve. $5 + 2(3 - 7x) = 11$ 0

68
Solve. $8x - 3(4 + 2x) = -6$ 3

69
Solve. $7x - 3 = 2x + 12$ 3

70
Solve. $3(2x - 5) + 7 = x + 2$ — 2

71
Solve. $19 + 2(2x - 5) + 21 = 5x + 6$ — 24

72
Solve. $3x - 7 + 5x = x + 7$ — 2

73
Solve. $x - 19 = 4(x + 2)$ — -9

74
Solve. $-30 + 3x = -2x$ — 6

75
Solve. $3x + 4 = 5(x - 4) + 2$ — 11

FEEDBACK UNIT 3

This quiz reviews the preceding unit. Answers are at the back of the book.

Find a solution for each of the following equations.

1. $x + 7 = -3$
2. $x - 4 = -1$
3. $5x = -40$
4. $-7x = 28$
5. $4x + 7 = 27$
6. $3x - 6 = 18$
7. $5x + 2 = x - 10$
8. $4 - 3x = 10 - x$
9. $5(2x + 4) - 17 = -7$
10. $4(x - 2) - 3x = 12$
11. $3(x - 3) - (x + 6) = 9x + 20$
12. $7x - 2(5x - 9) = 35 - 2(9 + x)$

UNIT 4: CHECKING SOLUTIONS

The following mathematical terms are crucial to an understanding of this unit.

Truth set
Check a solution
Empty set

1
The solution of an equation will make it a true statement. For this reason, the solution of an equation is often placed in braces and is the **truth set** for the equation. Find the truth set of $6(3x - 1) + 17 = 8x - 9 + 5x$.

{-4}

2
Find the truth set of $14x - 3 = 9x + 42$ by placing its solution in braces.

{9}

3
Find the truth set of $3(2 - x) + 9 = 2(3x + 1) - 5$.

{2}

4
Find the truth set of $5(2x - 7) = 3x - 35$.

{0}

5
Find the truth set of $-8x + 3 = 3x + 47$.

{-4}

6
Find the truth set of $7(2x - 5) + 31 = 17x + 8$.

{-4}

7
Find the truth set of $8x - 3 = 7x + 15$.

{18}

8
The truth set of $3x = 10$ is the **empty set**, { }, because there is no integer that can be multiplied by 3 to give 10. Find the truth set of $4x = -11$.

{ }

9
Find the truth set of $3(2x - 7) - (x - 6) = 6x - 16$.

{1}

10
Find the truth set of $5 + 2x = 3(x - 1)$. {8}

11
Find the truth set of $3(x - 2) + (2x + 5) = 9$. {2}

12
Find the truth set of $-27 + x = 4x + 3$. {-10}

13
Find the truth set of $3x - 4 = 6$. { }

14
Find the truth set of $9 - x = 3x + 9$. {0}

15
Find the truth set of $3(x - 5) = 15 - 7x$. {3}

16
The equation $x - 7 = 3$ has {10} as its truth set.
This means that when 10 replaces x in the equation
a true statement is obtained. It also means that when
any integer other than 10 replaces x a ________
statement is obtained. false

17
To **check** -6 as a possible solution for $5x - 3 = 3x - 9$
replace x by -6 and see whether the statement obtained
is true.

$$5x - 3 = 3x - 9$$

when $x = -6$

$$5(-6) - 3 = 3(-6) - 9$$
$$-30 - 3 = -18 - 9$$
$$-33 = -27$$

Since $-33 = -27$ is false, -6 _______(is, is not) in the truth
set of $5x - 3 = 3x - 9$. is not

18
To check 7 as a possible solution for $3(x - 5) = x - 1$
replace x by 7 and see whether the statement obtained is true.

$$3(x - 5) = x - 1$$

when $x = 7$

$$3(7 - 5) = 7 - 1$$
$$3(2) = 6$$
$$6 = 6$$

Since $6 = 6$ is true, 7 _______(is, is not) in the truth set of
$3(x - 5) = x - 1$. is

19

To check -2 as a solution for $-2(3x - 1) + 5 = -(x - 6) + 11$, the following steps are used:

$$-2(3x - 1) + 5 = -(x - 6) + 11$$

when $x = -2$ $$-2(3[-2] - 1) + 5 = -1(-2 - 6) + 11$$

$$-2(-6 - 1) + 5 = -1(-8) + 11$$

$$-2(-7) + 5 = 8 + 11$$

$$14 + 5 = 19$$

$$19 = 19$$

Therefore, -2 _______(is, is not) in the truth set of $-2(3x - 1) + 5 = -(x - 6) + 11$.

is

20

Check 6 as a solution for $2(x - 2) - 7 = x - 5$.

$2(6 - 2) - 7 = 6 - 5$
$2(4) - 7 = 1$
$8 - 7 = 1$
$1 = 1$ is true.
6 checks.

21

Check 2 as a solution for $-3x + 7 = 2x - 8$.

$-6 + 7 = 4 - 8$
$1 = -4$ is false.
2 doesn't check.

22

Check -4 as a solution for $-4x - 7 = x + 13$.

$16 - 7 = -4 + 13$
$9 = 9$ is true.
-4 checks.

23

Find and check the solution. $3(2x - 5) = -(x - 6)$

$x = 3$
$3(6 - 5) = -1(3 - 6)$
$3(1) = -1(-3)$
$3 = 3$ is true.

24

Find and check the solution. $5(2x + 3) = -2(x + 7) + 5$

$x = -2$
$5(-4 + 3) = -2(5) + 5$
$5(-1) = -10 + 5$
$-5 = -5$ is true.

25

Find and check the solution. $-(x - 3) + 4 = 2(x - 1)$

$x = 3$
$-(3 - 3) + 4 = 2(3 - 1)$
$-1(0) + 4 = 2(2)$
$4 = 4$ is true.

Feedback Unit 4

This quiz reviews the preceding unit. Answers are at the back of the book.

Find the truth set and show a check for each of the following equations.

1. $5x - 3 = 17$
2. $4x + 7 = x - 20$
3. $-2(x + 4) + 9 = x - 5$
4. $8 - 3(x - 5) = 2(3x - 4) + 4$
5. $-x + 2(x - 3) = 5(x + 4) - 6$

Unit 5: Applications

In this Applications Section, the format of the text has been altered. Answers for the problems appear beneath them rather than in the right-hand column. Your studying emphasis should be on learning the best procedures to follow with word problems.

1

There is a need to learn how to translate sentences into equations.

A number plus 13 is 25.	translates as	$N + 13 = 25$
15 more than a number equals 37.	translates as	$N + 15 = 37$
A number increased by 5 has a sum of 41.	translates as	$N + 5 = 41$

Write a translation for: A number increased by 3 gives 19.

Answer: The phrase "a number increased by 3" translates as $N + 3$. The verb "gives" translates as =. Consequently, the sentence translates as $N + 3 = 19$.

2

The following three examples are sentences involving subtraction. In each example, the sentence has been translated into an equation.

7 subtracted from a number equals 3.	translates as	$N - 7 = 3$
A number decreased by 8 is 23.	translates as	$N - 8 = 23$
12 diminished by a number gives 9.	translates as	$12 - N = 9$

Write a translation for: 9 less than a number is 16.

Answer: The phrase "9 less than a number" translates as $N - 9$. The verb "is" translates as =. Consequently, the sentence translates as $N - 9 = 16$.

3

The two examples below are sentences involving multiplication.

A number times 13 equals 26.	translates as	$13N = 26$
The product of a number and 8 is 24.	translates as	$8N = 24$

Write a translation for: A number increased threefold is 45.

Answer: The phrase "A number increased threefold" translates as 3N. The verb "is" translates as =. Consequently, the sentence translates as $3N = 45$.

4

A sentence can involve both addition and multiplication. Two examples are shown below.

Twice a number plus 7 is 23.	translates as	$2N + 7 = 23$
14 more than 3 times a number equals 35.	translates as	$3N + 14 = 35$

Write a translation for: 13 plus 5 times a number gives 28.

Answer:	13	plus	5 times a number	gives	28
	13	+	5N	=	28

The sentence may be translated as $13 + 5N = 28$ or $5N + 13 = 28$.

5

A sentence can involve both subtraction and multiplication.
Two examples are shown below.

9 less than twice a number equals 19.	translates as	$2N - 9 = 19$
Decreasing 5 times a number by 4 gives 31.	translates as	$5N - 4 = 31$

Write a translation for: 6 less than the product of a number and 7 is 22.

Answer:	6	less	product of a number and 7	is	22
	7N	–	6	=	22

The sentence may be translated as $7n - 6 = 22$.

FEEDBACK UNIT 5 FOR APPLICATIONS

1. Write an equation for: A number increased by 11 equals 48.

2. Write an equation for: 15 less than a number is 17.

3. Write an equation for: 7 multiplied by a number gives 56.

4. Write an equation for: 19 more than twice a number equals 27.

5. Write an equation for: 4 less than the product of a number and 9 is 67.

SUMMARY FOR SOLVING EQUATIONS WITH THE INTEGERS

The following mathematical terms are crucial to an understanding of this chapter.

Equation	Statement
Left side of an equation	Both sides of an equation
Equivalent equations	Solve an equation
Truth set	Empty set
Solve	Solution
Check a solution	

In Chapter 6 a method of finding truth sets for equations involving integers was presented. Equivalent equations were generated by adding opposites to both sides of an equality. By this process, new, easier equations can be created.

Some equations in this chapter required simplification of the expression on the left and/or right side of the equals sign before any other process should be used in finding a truth set.

Unit 4 presented the method for checking an equation's proposed solution. To check an equation, the proposed solution is substituted for the variable in the original equation. If the result is a true mathematical statement, the solution checks.

CHAPTER 6 MASTERY TEST

The following questions test the objectives of Chapter 6. Answers are at the back of the book. The number in parentheses which follows each problem indicates the unit in which it can be learned.

Solve each of the following equations.

1. $x + 9 = 3$ (1)
2. $x - 7 = 11$ (1)
3. $6y = -54$ (1)
4. $-7y = -14$ (1)
5. $5x + 4 = 29$ (2)
6. $8x - 6 = x + 8$ (2)
7. $3x + 4 = 9 - 2x$ (2)
8. $7 - 4x = 3 - 2x$ (2)
9. $8 + 5x = 3x - 22$ (2)
10. $7x - 9 = 3x + 51$ (2)
11. $2x - 8 = 9x - 57$ (2)
12. $6x - 13 = 31 + 7x$ (2)
13. $8(3x - 2) - 11x = 5x + 24$ (2)
14. $9 - 3(5x - 7) = 6(7 - x) - 3x$ (3)
15. $2(3x - 5) + 6 = 14$ (3)
16. $4(x - 6) + 3x = 25$ (3)

Find the truth set and show a check for each of the following equations.

17. $7x - 5 = 16$ (4)
18. $8x - 3 = 4x - 19$ (4)
19. $3(x - 7) = -x + 7(x - 3)$ (4)
20. $-x + 4 + 2x = 2(3x - 1) + 1$ (4)
21. Write an equation for:
    17 plus 6 times a number gives 65. (5)
22. Write an equation for: 9 less than the product of a number and 6 is 15. (5)

Chapter 7 Objectives

The following problems illustrate the objectives of this chapter. At this time you are not expected to know how to do these problems. However, if all of these problems are thoroughly understood, proceed directly to the Chapter 7 Mastery Test. The number in parentheses which follows each problem indicates the unit in which it can be learned.

1. Which of the following is not a rational number? (1)
   $\frac{4}{5}$ $\frac{-7}{8}$ $\frac{0}{9}$ $\frac{-3}{0}$ $\frac{47}{-2}$
2. The bar in a fraction $\frac{n}{d}$ indicates division. Any integer may replace n and any integer, except ______, may replace d. (1)
3. Is $\frac{-3}{5}$ equal to $\frac{3}{-5}$? (2)
4. Is $\frac{0}{7}$ equal to $\frac{0}{-8}$? (2)
5. Simplify. $\frac{-9}{12}$ (3)
6. Simplify. $\frac{-6}{-9}$ (3)
7. Evaluate. $\frac{3}{8} \cdot \frac{-7}{5}$ (4)
8. Evaluate. $0 \cdot \frac{-6}{7}$ (4)
9. Evaluate. $\frac{-3}{7} \cdot \frac{-2}{5}$ (4)
10. Evaluate. $\frac{13}{4} \cdot \frac{4}{13}$ (4)
11. Evaluate. $\frac{-27}{10} \cdot \frac{12}{25}$ (4)
12. Find the highest common factor of 40 and 16. (5)
13. Find the least common multiple of 8 and 20. (5)
14. The opposite of -9 is ______. (6)
15. Evaluate. $\frac{-3}{10} - \frac{7}{5}$ (6)
16. Evaluate. $\frac{8}{15} - \frac{-2}{75}$ (6)
17. Evaluate. $\frac{5}{24} + \frac{7}{40}$ (6)
18. The reciprocal of -5 is ______. (7)
19. Evaluate. $\frac{\frac{-2}{3}}{\frac{3}{7}}$ (7)
20. Evaluate. $\frac{5}{\frac{-5}{6}}$ (7)

CHAPTER 7
THE ARITHMETIC OF THE RATIONAL NUMBERS

UNIT 1: DEFINING RATIONAL NUMBERS

The following mathematical terms are crucial to an understanding of this unit.

Rational numbers	Ratio
Mixed numbers	Improper fraction
Numerator	Denominator

The equation $3x = 2$ cannot be made into a true statement by any integer, because there is no integer that can be multiplied by 3 to give a product of 2.

In this unit a new set of numbers will be studied to provide numbers that will make equations such as $2x = 5$ true statements. This new set of numbers is called the set of **rational numbers**.

1

$15 \div 3 =$ ______. 5

2

$18 \div 6 =$ ______. 3

3

$20 \div 5 =$ ______. 4

4

The division expression $12 \div 3$ can be written as the **ratio** $\frac{12}{3}$. Write the division expression $14 \div 7$ as a ratio. $\frac{14}{7}$

5

The division expression $25 \div 3$ can be written as the ratio $\frac{25}{3}$. Write $20 \div 9$ as a ratio. $\frac{20}{9}$

6

$17 \div 17$ can be written as $\frac{17}{17}$. Write $-21 \div -21$ as a ratio. $\frac{-21}{-21}$

7

$-2 \div 5$ can be written as $\frac{-2}{5}$. Write $-5 \div 8$ as a ratio. $\frac{-5}{8}$

8

$41 \div 73$ can be written as $\frac{41}{73}$. $25 \div 6$ can be written as ______. $\frac{25}{6}$

In arithmetic $\frac{13}{11}$ is often written as the **mixed number** $1\frac{2}{11}$, but in algebra the **improper fraction** $\frac{13}{11}$ is preferred. Mixed numbers like $1\frac{2}{11}$ will not be used in this text.

9

$2 \div 5$ can be written as $\frac{2}{5}$. $7 \div 15$ can be written as ______.

$\frac{7}{15}$

10

$15 \div -7$ can be written as $\frac{15}{-7}$. $4 \div -11$ can be written as ______.

$\frac{4}{-11}$

11

$3 \div 9$ can be written as ______.

$\frac{3}{9}$

12

$\frac{6}{2} = 3$ is a true statement because $3 \cdot 2 = 6$ is true.
Is $\frac{10}{2} = 5$ a true statement?

Yes (because $5 \cdot 2 = 10$)

13

$\frac{12}{3} = 4$ is a true statement because $4 \cdot 3 = 12$ is true.
Is $\frac{20}{4} = 5$ a true statement?

Yes (because $5 \cdot 4 = 20$)

14

$\frac{-15}{3} = -5$ is a true statement because $-5 \cdot 3 = -15$ is true.
Is $\frac{-24}{6} = -4$ a true statement?

Yes (because $-4 \cdot 6 = -24$)

15

$\frac{-16}{-8} = 2$ is a true statement because $2 \cdot -8 = -16$ is true.
Is $\frac{-30}{-3} = 10$ a true statement?

Yes (because $10 \cdot -3 = -30$)

16

$\frac{5}{2} = 3$ is a false statement because $3 \cdot 2 = 5$ is false.
Is $\frac{6}{12} = 2$ a true statement?

No ($2 \cdot 12 = 6$ is false)

17

$\frac{15}{7} = 2$ is a false statement because $2 \cdot 7 = 15$ is false.
Is $\frac{19}{9} = 2$ a true statement?

No ($2 \cdot 9 = 19$ is false)

18

$\frac{9}{4}$ is not equal to any integer because there is no integer that can be multiplied by 4 to obtain 9.
Is $\frac{10}{3}$ equal to any integer?

No

19

$\frac{2}{5}$ is not equal to any integer because the blank in 5 • ______ = 2 cannot be filled by any integer to give a true statement. Is $\frac{4}{7}$ equal to any integer?

No

20

Is $\frac{3}{4} = 2$ a true statement?

No

21

Is there any integer that is equal to $\frac{3}{4}$?

No

22

Is $\frac{8}{3} = 3$ a true statement?

No

23

Is there any integer equal to $\frac{8}{3}$?

No

24

Is $\frac{5}{10} = 2$ a true statement?

No (2 • 10 = 5 is false)

25

Is there any integer equal to $\frac{5}{10}$?

No

26

There is no integer replacement for x in 2x = 7 that makes a true statement. Is there any integer replacement for x in 3x = 4 that makes a true statement?

No

27

Is there any integer replacement for x in 5x = 1 that makes a true statement?

No

28

$\frac{3}{4}$ is not an integer. $\frac{3}{4}$ is the number that will make 4x = 3 true. Is there an integer that will make 7x = 6 true?

No

29

$\frac{6}{7}$ is not an integer. $\frac{6}{7}$ is the number that will make 7x = 6 true. Is there an integer that will make 0x = 5 true?

No

30

To make 0x = 5 true, it would be necessary to find a number that could be multiplied by zero to give 5. Any number multiplied by zero is equal to ______.

zero (0)

31

$\frac{5}{9}$ is a **rational number**. For the open expression $\frac{x}{y}$, if x is replaced by any integer and y is replaced by any integer except zero, the result is a rational number. Is $\frac{3}{7}$ a rational number?

Yes

32

For the open expression $\frac{x}{y}$, x can be replaced by any integer and y can be replaced by any integer except zero. The result from such replacements is a rational number. Replace x by 2 and y by 9 in $\frac{x}{y}$. What is the rational number obtained?

$\frac{2}{9}$

33

Numbers such as $\frac{5}{6}$, $\frac{-71}{9}$, $\frac{8}{3}$, and $\frac{4}{-7}$ are rational numbers. Is $\frac{3}{4}$ a rational number?

Yes

34

Is $\frac{14}{25}$ a rational number?

Yes

35

Is $\frac{-16}{9}$ a rational number?

Yes

36

Is $\frac{46}{-3}$ a rational number?

Yes

37

For the open expression $\frac{x}{y}$, any integer may replace x and any integer except zero may replace y. $\frac{8}{0}$ is not a rational number. Is $\frac{13}{0}$ a rational number?

No

38

Is $\frac{-15}{0}$ a rational number?

No

$\frac{13}{0}$, $\frac{-5}{0}$, and $\frac{0}{0}$ are not rational numbers because division by zero is impossible. Any expression of the form $\frac{x}{0}$ cannot be a rational number because zero cannot be a divisor.

$\frac{4}{0}$ does not equal 4 because $4 \cdot 0$ is not equal to 4. Similarly, $\frac{4}{0}$ does not equal zero, because $0 \cdot 0$ does not equal 4.

39
For the open expression $\frac{x}{y}$, any integer may replace x and any integer except zero may replace y. $\frac{0}{3}$ is a rational number. Is $\frac{0}{8}$ a rational number?

Yes, $8 \cdot 0 = 0$

40
Is $\frac{0}{6}$ a rational number?

Yes, $6 \cdot 0 = 0$

41
Is $\frac{0}{-4}$ a rational number?

Yes, $-4 \cdot 0 = 0$

42
$\frac{0}{8}$ is a rational number. $\frac{8}{0}$ is not a rational number.
$\frac{15}{0}$ ______ (is, is not) a rational number.

is not

43
For the rational number $\frac{5}{9}$, the integer above the bar is called the **numerator**. For the rational number $\frac{5}{9}$, 5 is called the ______.

numerator

44
4 is the numerator of $\frac{4}{7}$ because 4 is above the bar.
For the rational number $\frac{6}{11}$, 6 is the ______.

numerator

45
For the rational number $\frac{3}{10}$, 3 is the ______ because it is above the bar.

numerator

46

For the rational number $\frac{9}{5}$, 9 is the ______. | numerator

47

For the rational number $\frac{4}{7}$, the integer below the bar is called the **denominator**. For the rational number $\frac{4}{7}$, 7 is the ______. | denominator

48

5 is the denominator of $\frac{-2}{5}$ because 5 is below the bar. For the rational number $\frac{-17}{8}$, 8 is the ______. | denominator

49

For the rational number $\frac{-3}{5}$, 5 is the ______ because 5 is below the bar. | denominator

50

For the rational number $\frac{14}{19}$, 19 is the denominator and 14 is the ______. | numerator

51

For the rational number $\frac{-2}{-5}$, -2 is the numerator and -5 is the ______. | denominator

52

For the rational number $\frac{-19}{13}$, the numerator is ______ and the denominator is ______. | -19, 13

53

The numerator of a rational number may be any integer. The denominator of a rational number may be any integer except ______. | zero

FEEDBACK UNIT 1

This quiz reviews the preceding unit. Answers are at the back of the book.

1. Write the following division problems as ratios.
   a. $5 \div 13$ b. $15 \div -7$ c. $-12 \div -4$
   d. $-9 \div 2$ e. $0 \div 5$
2. Is $\frac{-8}{4} = -4$?
3. Is $\frac{12}{-3} = -4$?
4. Is $\frac{-15}{-15} = 1$?
5. The denominator of $\frac{8}{15}$ is ______.
6. The numerator of $\frac{-7}{12}$ is ______.
7. Is $\frac{-27}{11}$ a rational number?
8. Is $\frac{0}{4}$ a rational number?
9. Is $\frac{-5}{-9}$ a rational number?
10. Is $\frac{3}{0}$ a rational number?

UNIT 2: EQUAL RATIONAL NUMBERS

The following mathematical terms are crucial to an understanding of this unit.

Integers as rational numbers Equal rational numbers

Every counting number, {1, 2, 3, . . . }, is also an integer, {. . . , -2, -1, 0, 1, 2, . . . }. In this unit it is shown that every integer is in the set of rational numbers.

1
The integer 5 can be written as the rational number $\frac{5}{1}$.
Write the integer 9 as a rational number.

$\frac{9}{1}$

2
Every integer is a rational number. Write the integer 37 as a rational number.

$\frac{37}{1}$

3
The integer -12 can be written as the rational number $\frac{-12}{1}$. Write -15 as a rational number.

$\frac{-15}{1}$

4
Write the integer -22 as a rational number.

$\frac{-22}{1}$

5
To write any integer as a rational number, the integer is written as the numerator and positive one (1) is used as the denominator. Write -5 as a rational number.

$\frac{-5}{1}$

6
$\frac{3}{4}, \frac{6}{8}, \frac{9}{12}$, etc., are names for the same rational number. $\frac{2}{6}$ can be written more simply as $\frac{1}{3}$. Therefore, $\frac{2}{6} = \frac{1}{3}$. Is $2 \cdot 3 = 6 \cdot 1$ a true statement?

Yes

7
$\frac{9}{12}$ can be written more simply as $\frac{3}{4}$. Therefore, $\frac{9}{12} = \frac{3}{4}$. Is $9 \cdot 4 = 12 \cdot 3$ a true statement?

Yes

8
$\frac{4}{6}$ can be written more simply as $\frac{2}{3}$. Therefore, $\frac{4}{6} = \frac{2}{3}$. Is $4 \cdot 3 = 6 \cdot 2$ a true statement?

Yes

9
$\frac{5}{10} = \frac{1}{2}$ is true because $5 \cdot 2 = 10 \cdot 1$ is true.
$\frac{4}{6} = \frac{2}{3}$ is true because $4 \cdot 3 = 6 \cdot$ ______ is true.

2

10
To decide whether or not $\frac{3}{5} = \frac{6}{10}$ is true, use the statement $3 \cdot 10 = 5 \cdot 6$. If $3 \cdot 10 = 5 \cdot 6$ is true, then $\frac{3}{5} = \frac{6}{10}$. To decide whether or not $\frac{4}{7} = \frac{12}{21}$ is a true statement, use the statement $4 \cdot 21 = 7 \cdot$ ______.

12

To decide whether or not **two rational numbers are equal:**

1. Multiply the first numerator by the second denominator.
2. Multiply the first denominator by the second numerator.
3. When the products are the same, then the rational numbers are equal.

$\frac{6}{12} = \frac{1}{2}$ is true because $6 \cdot 2 = 12 \cdot 1$ is true.

$\frac{7}{5} = \frac{14}{9}$ is false because $7 \cdot 9 = 5 \cdot 14$ is false.

11

To decide if $\frac{2}{5} = \frac{4}{10}$ is true, multiply 2 by 10 and ______ by ______.

5 by 4

12

To decide if $\frac{5}{10} = \frac{1}{2}$ is true, compare $5 \cdot 2$ with $10 \cdot 1$. Since $5 \cdot 2 = 10 \cdot 1$ is true, then $\frac{5}{10} = \frac{1}{2}$ is ______ (true or false).

true

13

$\frac{3}{5} = \frac{6}{10}$ is true because $3 \cdot 10 = 5 \cdot 6$ is true.

$\frac{8}{10} = \frac{4}{5}$ is true because __________ is true.

$8 \cdot 5 = 10 \cdot 4$

14

$\frac{4}{5} = \frac{8}{10}$ is true because __________ is true.

$4 \cdot 10 = 5 \cdot 8$

15

$\frac{5}{10} = \frac{1}{3}$ is false because $5 \cdot 3 = 10 \cdot 1$ is false.

Is $\frac{5}{8} = \frac{2}{3}$ true?

No, 15 is not equal to 16.

16

Is $\frac{5}{9} = \frac{20}{36}$ a true statement?

Yes, 180 = 180

17

Is $\frac{7}{12} = \frac{9}{20}$ a true statement?

No

18

Is $\frac{14}{42} = \frac{1}{2}$ a true statement?

No

19

Is $\frac{9}{15} = \frac{2}{3}$ a true statement?

No

20
Is $\frac{38}{57} = \frac{2}{3}$ a true statement?

Yes

21
Is $\frac{2}{13} = \frac{10}{65}$ a true statement?

Yes

22
Is $\frac{3}{12} = \frac{18}{72}$ a true statement?

Yes

23
Is $\frac{0}{5} = \frac{5}{0}$ a true statement?

No

24
$\frac{-5}{10} = \frac{-1}{2}$ is true because $-5 \bullet 2 = 10 \bullet -1$ is true.
Is $\frac{-4}{16} = \frac{-1}{4}$ true?

Yes

25
Is $\frac{-6}{10} = \frac{-9}{15}$ true?

Yes

26
Is $\frac{5}{9} = \frac{-5}{9}$ true?

No, 45 is not equal to -45

27
Is $\frac{3}{-4} = \frac{-3}{4}$ true?

Yes, 12 = 12

28
When a rational number is the ratio $(\frac{x}{y})$ of one negative integer and one positive integer, it should be written with a negative numerator and a positive denominator. For example, $\frac{-3}{4}$ is equal to $\frac{3}{-4}$ and the preferred way of writing it is _______.

$\frac{-3}{4}$ because the numerator is negative and the denominator is positive.

29
Is $\frac{-2}{5} = \frac{2}{-5}$ true?

Yes, 10 = 10

30
Since $\frac{-2}{5}$ and $\frac{2}{-5}$ are equal and it is preferred that the denominator be positive, the preferred way of writing the rational number is _______.

$\frac{-2}{5}$

31
Is $\frac{-3}{5} = \frac{3}{5}$ true?

No, -15 is not equal to 15

32
Is $\frac{-5}{6} = \frac{5}{-6}$ true?

Yes, 30 = 30

33
Since $\frac{-5}{6}$ and $\frac{5}{-6}$ are equal, the preferred way of writing the rational number is _______.

$\frac{-5}{6}$

34
Is $\frac{7}{-3} = \frac{-7}{3}$ true?

Yes, 21 = 21

35
Since $\frac{-7}{3}$ and $\frac{7}{-3}$ are equal, the preferred way of writing the rational number is _______.

$\frac{-7}{3}$

36
Is $\frac{-4}{9} = \frac{4}{-9}$ true?

Yes, 36 = 36

37
Is $\frac{2}{7} = \frac{-2}{7}$ true?

No, 14 is not equal to -14

38
When a rational number is the ratio of two counting numbers, can it be equal to the ratio of a negative integer and a counting number?

No, like $\frac{2}{7}$ and $\frac{-2}{7}$ the multiplications are unequal

39
Is $\frac{0}{5} = \frac{0}{-2}$ true?

Yes, $0 \cdot -2 = 5 \cdot 0$

40
Any rational number of the form $\frac{0}{y}$ where y is a non-zero integer will be equal to 0. Is $\frac{0}{8} = 0$ true?

Yes

41
Each of the rational numbers listed below is equal to 0.

$\frac{0}{9}, \frac{0}{-5}, \frac{0}{14}, \frac{0}{7}, \frac{0}{-3}, \frac{0}{45}$

When the numerator is 0 and the denominator is not zero, the rational number is _____.

0

42
Is $\frac{-12}{-6} = 2$ true?

Yes, 2 is $\frac{2}{1}$ and -12 = -12

43
Is $\frac{-2}{-8} = \frac{1}{4}$ true?

Yes, $-2 \cdot 4 = -8 \cdot 1$

44
Is $\frac{-20}{-5} = 4$ true?

Yes, $-20 = -20$

45
Is $\frac{-25}{-40} = \frac{5}{8}$ true?

Yes, $-200 = -200$

46
Is $\frac{9}{-27} = \frac{-1}{3}$ true?

Yes, $27 = 27$

47
Since $\frac{9}{-27}$ equals $\frac{-1}{3}$, which one is preferred?

$\frac{-1}{3}$, denominator is positive.

48
Is $\frac{-5}{-7} = \frac{5}{7}$ true?

Yes, $-35 = -35$

49
Since $\frac{-5}{-7}$ equals $\frac{5}{7}$, which one is preferred?

$\frac{5}{7}$, denominator is positive.

50
Is $\frac{4}{11} = \frac{-4}{11}$ true?

No

51
Is $\frac{6}{5} = \frac{-6}{-5}$ true?

Yes

52
Since $\frac{6}{5}$ equals $\frac{-6}{-5}$, which one is preferred?

$\frac{6}{5}$, denominator is positive.

53
Below are shown three pairs of equal rational numbers where the numerators and denominators have opposite signs, positive or negative.

$$\frac{-8}{9} = \frac{8}{-9} \qquad \frac{7}{-10} = \frac{-7}{10} \qquad \frac{4}{-3} = \frac{-4}{3}$$

For each pair, the preferred rational number has a positive ______________.

denominator

54
Write $\frac{4}{-9}$ with a positive denominator.

$\frac{-4}{9}$

55
Write $\frac{9}{-16}$ with a positive denominator.

$\frac{-9}{16}$

56
Write $\frac{11}{-5}$ with a positive denominator.

$\frac{-11}{5}$

57
Write $\frac{7}{-15}$ with a positive denominator.

$\frac{-7}{15}$

58
Below are shown three pairs of equal rational numbers where the numerators and denominators have the same signs, positive or negative.

$\frac{3}{13} = \frac{-3}{-13} \qquad \frac{-9}{-10} = \frac{9}{10} \qquad \frac{-7}{-5} = \frac{7}{5}$

For each pair, the preferred rational number has a positive ____________.

denominator

59
Write $\frac{-9}{-8}$ with a positive denominator.

$\frac{9}{8}$

60
Write $\frac{-7}{-15}$ with a positive denominator.

$\frac{7}{15}$

61
Write $\frac{-9}{-4}$ with a positive denominator.

$\frac{9}{4}$

62
Write $\frac{17}{-6}$ with a positive denominator.

$\frac{-17}{6}$

63
Write $\frac{-13}{-5}$ with a positive denominator.

$\frac{13}{5}$

64
Write $\frac{2}{-5}$ with a positive denominator.

$\frac{-2}{5}$

FEEDBACK UNIT 2

This quiz reviews the preceding unit. Answers are at the back of the book.

1. Write 2 as a rational number.
2. Write -15 as a rational number.
3. Is $\frac{2}{3} = \frac{8}{12}$ true?
4. Is $\frac{-5}{9} = \frac{5}{-9}$ true?
5. Is $\frac{-2}{-5} = \frac{-2}{5}$ true?
6. Is $\frac{-10}{-2} = 5$ true?
7. Is $\frac{0}{8} = \frac{-1}{8}$ true?
8. Is $\frac{0}{16} = \frac{0}{-7}$ true?
9. Is $\frac{-14}{7} = 2$ true?
10. Is $\frac{-4}{-7} = \frac{4}{7}$ true?
11. Write $\frac{-13}{-15}$ with a positive denominator.
12. Write $\frac{4}{-11}$ with a positive denominator.

UNIT 3: SIMPLIFYING RATIONAL NUMBERS

The following mathematical terms are crucial to an understanding of this unit.

Factors
Common factor
Simplifying rational numbers
Highest common factor, HCF
Simplest name for a rational number

1
The **factors** of 12 are: 1, 2, 3, 4, 6, 12.
The factors of 18 are: ______________.

1, 2, 3, 6, 9, 18

2
The integers 1, 2, 3, and 6 are factors of both 12 and 18. The **highest common factor, HCF,** for 12 and 18 is ______ because it is the largest factor they have in common.

6

3
The factors of 10 are: 1, 2, 5, 10.
The factors of 15 are: ______________.

1, 3, 5, 15

4
The integers 1 and 5 are **common factors** of 10 and 15. The highest common factor, HCF, for 10 and 15 is ______.

5

5
What is the highest common factor, HCF, of 12 and 15?

3, only 1 and 3 are common factors.

6
What is the HCF of 14 and 21?

7, only 1 and 7 are common factors.

7
What is the HCF of 8 and 20?

4, the common factors are: 1, 2, 4

8
What is the HCF of 10 and 30?

10, the common factors are: 1, 2, 5, 10

9
The factors of 6 are: 1, 2, 3, 6. The factors of 7 are: 1, 7. What is the HCF of 6 and 7?

1, the only common factor is 1.

10
What is the highest common factor of 3 and 8?

1

11
What is the highest common factor of 4 and 6?

2

12
What is the HCF of 20 and 30?

10

13
What is the HCF of 15 and 18?

3

14
The rational numbers $\frac{2}{3}$ and $\frac{8}{12}$ are equal, but the simpler one is $\frac{2}{3}$. Four ways of showing the same rational number are: $\frac{1}{2}, \frac{2}{4}, \frac{3}{6}, \frac{4}{8}$. Which is the simplest one?

$\frac{1}{2}$

15
The rational numbers $\frac{-9}{12}$ and $\frac{-3}{4}$ are equal, but the simpler one is $\frac{-3}{4}$. Four ways of showing the same rational number are: $\frac{12}{30}, \frac{8}{20}, \frac{4}{10}, \frac{2}{5}$. Which is the simplest one?

$\frac{2}{5}$

16
Every rational number has an unlimited (infinite) number of names. The **simplest name** for $\frac{3}{6}$ is $\frac{1}{2}$, because 2 is the smallest possible positive denominator. The simplest name for $\frac{10}{15}$ is $\frac{2}{3}$, because 3 is the ______ positive denominator.

smallest

17
To find the simplest name for $\frac{14}{49}$, it is necessary to find the HCF for 14 and 49. What is the highest common factor for 14 and 49?

7

18
7 is the HCF for 14 and 49. To **simplify** $\frac{14}{49}$, divide both 14 and 49 by 7.

$$\frac{14}{49} = \frac{14 \div 7}{49 \div 7} = \frac{2}{7}$$

The simplest name for $\frac{14}{49}$ is ______.

$\frac{2}{7}$

19
To simplify $\frac{8}{20}$, use the HCF of 8 and 20.

$$\frac{8}{20} = \frac{8 \div 4}{20 \div 4} = \frac{2}{5}$$

The simplest name for $\frac{8}{20}$ is ______.

$\frac{2}{5}$

20
To simplify $\frac{4}{14}$, use the HCF of 4 and 14.

$$\frac{4}{14} = \frac{4 \div 2}{14 \div 2} = \frac{2}{7}$$

The simplest name for $\frac{4}{14}$ is ______.

$\frac{2}{7}$

21
To simplify $\frac{6}{16}$, use the HCF of 6 and 16.

$$\frac{6}{16} = \frac{6 \div 2}{16 \div 2} = \frac{3}{8}$$

The simplest name for $\frac{6}{16}$ is ______.

$\frac{3}{8}$

22
Simplify. $\frac{10}{4}$

$\frac{10 \div 2}{4 \div 2} = \frac{5}{2}$

23
Simplify. $\frac{15}{5}$

$\frac{15 \div 5}{5 \div 5} = \frac{3}{1} = 3$

24
Simplify. $\frac{4}{6}$

$\frac{2}{3}$

25
Simplify. $\frac{10}{25}$

$\frac{2}{5}$

26
Simplify. $\frac{21}{49}$

$\frac{3}{7}$

27
Simplify. $\frac{7}{28}$

$\frac{1}{4}$

28
Simplify. $\frac{9}{15}$

$\frac{3}{5}$

29
Simplify. $\frac{24}{8}$

3

30
Simplify. $\frac{30}{18}$

$\frac{5}{3}$

31
Simplify. $\frac{12}{20}$

$\frac{3}{5}$

32
To simplify $\frac{-6}{8}$, first determine that the HCF of -6 and 8 is 2.

$$\frac{-6}{8} = \frac{-6 \div 2}{8 \div 2} = \frac{-3}{4}$$

Simplify. $\frac{-10}{16}$

$\frac{-5}{8}$

33
Complete the simplification of $\frac{-15}{21}$:

$$\frac{-15}{21} = \frac{-15 \div 3}{21 \div 3} = \_\_\_\_\_\_.$$

$\frac{-5}{7}$

34
Complete the simplification of $\frac{-12}{3}$:

$$\frac{-12}{3} = \frac{-12 \div 3}{3 \div 3} = \_\_\_\_\_\_.$$

-4

35
Simplify. $\frac{-9}{6}$

$\frac{-3}{2}$

36
Simplify. $\frac{-7}{35}$

$\frac{-1}{5}$

37
Simplify. $\frac{12}{48}$

$\frac{1}{4}$

38
Simplify. $\frac{-8}{24}$

$\frac{-1}{3}$

Recall that any rational number with a negative denominator can be simplified so it has a positive denominator.

$\frac{-2}{-5}$ is simplified to $\frac{2}{5}$.

$\frac{6}{-7}$ is simplified to $\frac{-6}{7}$.

39
Simplify $\frac{4}{-8}$ by first replacing it with $\frac{-4}{8}$.

$\frac{-1}{2}$

40
Simplify $\frac{-20}{-10}$ by first replacing it with $\frac{20}{10}$.

2

41
Simplify. $\frac{-3}{-18}$

$\frac{1}{6}$

42
Simplify. $\frac{14}{-7}$

-2

43
Simplify. $\frac{12}{-14}$

$\frac{-6}{7}$

44
Simplify. $\frac{15}{20}$

$\frac{3}{4}$

FEEDBACK UNIT 3

This quiz reviews the preceding unit. Answers are at the back of the book.

1. Simplify. $\frac{15}{25}$
2. Simplify. $\frac{3}{12}$
3. Simplify. $\frac{18}{-20}$
4. Simplify. $\frac{16}{40}$
5. Simplify. $\frac{-6}{12}$
6. Simplify. $\frac{-20}{8}$
7. Simplify. $\frac{-6}{15}$
8. Simplify. $\frac{-20}{-35}$

UNIT 4: MULTIPLICATION OF RATIONAL NUMBERS

The following mathematical terms are crucial to an understanding of this unit.

Evaluating a multiplication   Cancelling

1
To **evaluate the multiplication** of two rationals:
1. Multiply the numerators.
2. Multiply the denominators.

Evaluate. $\frac{1}{2} \bullet \frac{3}{4}$

$\frac{3}{8}$

2
Complete the evaluation: $\frac{2}{3} \bullet \frac{5}{7} = \frac{2 \bullet 5}{3 \bullet 7} =$ ______.

$\frac{10}{21}$

3
Evaluate. $\frac{4}{9} \bullet \frac{5}{7}$

$\frac{20}{63}$

4
Evaluate. $\frac{3}{4} \bullet \frac{5}{7}$

$\frac{15}{28}$

5
Evaluate. $\frac{1}{5} \bullet \frac{2}{7}$

$\frac{2}{35}$

6
To evaluate $\frac{-3}{5} \bullet \frac{4}{11}$, the following steps are used:

$\frac{-3}{5} \bullet \frac{4}{11} = \frac{-3 \bullet 4}{5 \bullet 11} = \frac{-12}{55}$

Evaluate. $\frac{2}{7} \bullet \frac{-4}{3}$

$\frac{-8}{21}$

7
Complete the evaluation: $\frac{5}{8} \bullet \frac{-3}{7} = \frac{5 \bullet -3}{8 \bullet 7} =$ ______.

$\frac{-15}{56}$

8
Evaluate. $\frac{-2}{7} \bullet \frac{1}{5}$

$\frac{-2}{35}$

9
Evaluate. $\frac{7}{9} \cdot \frac{-2}{5}$

$\frac{-14}{45}$

10
Complete the evaluation: $\frac{-4}{13} \cdot 5 = \frac{-4}{13} \cdot \frac{5}{1} =$ ______.

$\frac{-20}{13}$

11
Evaluate $\frac{-2}{7} \cdot 3$ by first writing 3 as a rational number.

$\frac{-6}{7}$

12
Evaluate. $\frac{-4}{5} \cdot 6$

$\frac{-24}{5}$

13
Evaluate. $\frac{2}{7} \cdot -4$

$\frac{-8}{7}$

14
Evaluate. $5 \cdot \frac{-3}{14}$

$\frac{-15}{14}$

15
Evaluate. $\frac{4}{3} \cdot \frac{-14}{5}$

$\frac{-56}{15}$

16
Evaluate. $\frac{4}{7} \cdot \frac{-3}{5}$

$\frac{-12}{35}$

17
Any number times zero is zero.

$$\frac{-3}{8} \cdot 0 = \frac{-3}{8} \cdot \frac{0}{1} = \frac{-3 \cdot 0}{8 \cdot 1} = \frac{0}{8} = 0$$

Evaluate. $\frac{-7}{9} \cdot 0$

0

18
Evaluate. $\frac{4}{3} \cdot 0$

0

19
Evaluate. $\frac{-7}{16} \cdot 0$

0

20
Evaluate. $\frac{-5}{6} \cdot 0$

0

21
Evaluate. $0 \cdot \frac{4}{7}$

0

22
To evaluate $\frac{-3}{8} \bullet \frac{-1}{2}$, the following steps are used:

$$\frac{-3}{8} \bullet \frac{-1}{2} = \frac{-3 \bullet -1}{8 \bullet 2} = \frac{3}{16}$$

Evaluate. $\frac{-4}{5} \bullet \frac{-2}{7}$

$\frac{8}{35}$

23
Complete the evaluation: $\frac{-4}{7} \bullet \frac{-3}{5} = \frac{-4 \bullet -3}{7 \bullet 5} =$ ______.

$\frac{12}{35}$

24
Evaluate. $\frac{-9}{4} \bullet \frac{-3}{2}$

$\frac{27}{8}$

25
Evaluate. $\frac{-1}{2} \bullet \frac{-3}{5}$

$\frac{3}{10}$

26
Evaluate. $\frac{-9}{7} \bullet \frac{1}{4}$

$\frac{-9}{28}$

27
Any rational number multiplied by 1 produces identically the same rational number. Evaluate $\frac{5}{8} \bullet 1$.

$\frac{5}{8}$

28
Complete the evaluation: $\frac{-4}{17} \bullet 1 = \frac{-4}{17} \bullet \frac{1}{1} = \frac{-4 \bullet 1}{17 \bullet 1} =$ ______.

$\frac{-4}{17}$

29
Evaluate. $\frac{-3}{4} \bullet 1$

$\frac{-3}{4}$

30
Evaluate. $\frac{-5}{8} \bullet 0$

0

31
Evaluate. $\frac{5}{4} \bullet \frac{7}{3}$

$\frac{35}{12}$

32
The multiplication of integers is commutative and guarantees that $4 \bullet 3$ has the same evaluation as $3 \bullet 4$.

Does this guarantee that $\frac{3 \bullet 7}{4 \bullet 3}$ will be equal to $\frac{3 \bullet 7}{3 \bullet 4}$?

Yes

33

For the integers 5 and 7, the Commutative Law of Multiplication guarantees that $5 \cdot 7$ has the same evaluation as $7 \cdot 5$. Does this guarantee that $\frac{7 \cdot 9}{5 \cdot 7}$ will be equal to $\frac{7 \cdot 9}{7 \cdot 5}$?

Yes

34

The rational numbers $\frac{4 \cdot 9}{5 \cdot 4}$ and $\frac{4 \cdot 9}{4 \cdot 5}$ must be equal because $5 \cdot 4$ equals $4 \cdot 5$. Similarly, are $\frac{2 \cdot 7}{9 \cdot 2}$ and $\frac{2 \cdot 7}{2 \cdot 9}$ equal?

Yes

35

Because the multiplication of integers is commutative, the order of the denominators in the problem below can be reversed. This simplifies the multiplication.

$$\frac{2}{9} \cdot \frac{7}{2} = \frac{2 \cdot 7}{9 \cdot 2} = \frac{2 \cdot 7}{2 \cdot 9} = \frac{2}{2} \cdot \frac{7}{9} = 1 \cdot \frac{7}{9} = \frac{7}{9}$$

Use the fact that $\frac{4}{4}$ equals 1 to simplify the evaluation of $\frac{4}{7} \cdot \frac{9}{4}$.

$1 \cdot \frac{9}{7} = \frac{9}{7}$

36

Complete the evaluation: $\frac{8}{9} \cdot \frac{-7}{8} = \frac{8}{8} \cdot \frac{-7}{9} = 1 \cdot \frac{-7}{9} =$ ______.

$\frac{-7}{9}$

37

Complete the evaluation: $\frac{9}{8} \cdot \frac{5}{9} = \frac{9}{9} \cdot \frac{5}{8} = 1 \cdot \frac{5}{8} =$ ______.

$\frac{5}{8}$

38

The process shown in the last six frames is the basis for **cancelling** in the multiplication of rational numbers. Will $\frac{5}{3} \cdot \frac{3}{8}$ have the same evaluation as $\frac{3}{3} \cdot \frac{5}{8}$?

Yes

39

Cancelling is a useful simplification in the multiplication of rational numbers. Will $\frac{7}{9} \cdot \frac{-4}{7}$ have the same evaluation as $\frac{7}{7} \cdot \frac{-4}{9}$?

Yes

40

The example below shows the cancelling process as it is normally done. This example relies on the fact that 7 is a factor of both a numerator and a denominator.

$$\frac{2}{7} \bullet \frac{7}{9} = \frac{2}{\cancel{7}_1} \bullet \frac{\cancel{7}^1}{9} = \frac{2}{1} \bullet \frac{1}{9} = \frac{2}{9}$$

To evaluate $\frac{7}{6} \bullet \frac{6}{11}$, cancelling is possible because ____ is a factor of both a numerator and a denominator.

6

41

Complete the evaluation: $\frac{7}{6} \bullet \frac{6}{11} = \frac{7}{\cancel{6}_1} \bullet \frac{\cancel{6}^1}{11} = \frac{7}{1} \bullet \frac{1}{11} =$ ______.

$\frac{7}{11}$

42

Use cancellation to evaluate $\frac{4}{5} \bullet \frac{7}{4}$.

$\frac{1}{5} \bullet \frac{7}{1} = \frac{7}{5}$

43

Use cancellation to evaluate $\frac{2}{3} \bullet \frac{3}{5}$.

$\frac{2}{1} \bullet \frac{1}{5} = \frac{2}{5}$

44

Use cancellation to evaluate $\frac{5}{8} \bullet \frac{8}{11}$.

$\frac{5}{11}$

45

Use cancellation to evaluate $\frac{4}{7} \bullet \frac{3}{4}$.

$\frac{3}{7}$

46

Evaluate. $\frac{3}{7} \bullet \frac{4}{3}$

$\frac{4}{7}$

47

Evaluate. $\frac{4}{9} \bullet \frac{7}{4}$

$\frac{7}{9}$

48

Evaluate. $\frac{16}{19} \bullet \frac{7}{16}$

$\frac{7}{19}$

49

To evaluate $\frac{2}{5} \bullet \frac{-1}{2}$, the following steps are used:

$$\frac{2}{5} \bullet \frac{-1}{2} = \frac{\cancel{2}^1}{5} \bullet \frac{-1}{\cancel{2}_1} = \frac{1}{5} \bullet \frac{-1}{1} = \frac{-1}{5}$$

Evaluate. $\frac{7}{9} \bullet \frac{-2}{7}$

$\frac{-2}{9}$

50
Evaluate. $\frac{11}{13} \bullet \frac{-4}{11}$

$\frac{-4}{13}$

51
Evaluate. $\frac{-5}{13} \bullet \frac{13}{6}$

$\frac{-5}{6}$

52
Evaluate. $\frac{-3}{11} \bullet \frac{11}{14}$

$\frac{-3}{14}$

53
Evaluate. $\frac{14}{5} \bullet \frac{-3}{14}$

$\frac{-3}{5}$

54
The evaluation shown below uses the fact that $\frac{-6}{6} = \frac{-1}{1}$.

$$\frac{-6}{5} \bullet \frac{7}{6} = \frac{\overset{-1}{\cancel{-6}}}{5} \bullet \frac{7}{\underset{1}{\cancel{6}}} = \frac{-1}{5} \bullet \frac{7}{1} = \frac{-7}{5}$$

Evaluate. $\frac{-3}{8} \bullet \frac{5}{3}$

$\frac{-1}{8} \bullet \frac{5}{1} = \frac{-5}{8}$

55
Evaluate. $\frac{-5}{7} \bullet \frac{8}{5}$

$\frac{-8}{7}$

56
Evaluate. $\frac{-7}{12} \bullet \frac{5}{7}$

$\frac{-5}{12}$

57
Evaluate. $\frac{-2}{7} \bullet \frac{-9}{2}$

$\frac{-1}{7} \bullet \frac{-9}{1} = \frac{9}{7}$

58
Evaluate. $\frac{-3}{5} \bullet \frac{-4}{3}$

$\frac{4}{5}$

59
Evaluate. $\frac{-5}{9} \bullet \frac{-4}{5}$

$\frac{4}{9}$

60
Evaluate. $\frac{6}{13} \bullet \frac{-13}{11}$

$\frac{-6}{11}$

61
Evaluate. $\frac{15}{4} \bullet \frac{7}{15}$

$\frac{7}{4}$

62
Evaluate. $\frac{-3}{11} \bullet \frac{-7}{3}$

$\frac{7}{11}$

63
To use cancellation with $\frac{12}{5} \bullet \frac{7}{6}$ first find the highest common factor, HCF, of 12 and 6.

$$\frac{12}{5} \bullet \frac{7}{6} = \frac{\overset{2}{\cancel{12}}}{5} \bullet \frac{7}{\underset{1}{\cancel{6}}} = \frac{2}{5} \bullet \frac{7}{1} = \frac{14}{5}$$

Evaluate. $\frac{14}{3} \bullet \frac{5}{7}$

$\frac{2}{3} \bullet \frac{5}{1} = \frac{10}{3}$

64
Evaluate. $\frac{8}{5} \bullet \frac{7}{4}$

$\frac{2}{5} \bullet \frac{7}{1} = \frac{14}{5}$

65
Evaluate. $\frac{10}{11} \bullet \frac{3}{5}$

$\frac{2}{11} \bullet \frac{3}{1} = \frac{6}{11}$

66
Evaluate. $\frac{4}{5} \bullet \frac{7}{12}$

$\frac{1}{5} \bullet \frac{7}{3} = \frac{7}{15}$

67
Evaluate. $\frac{3}{20} \bullet \frac{5}{8}$

$\frac{3}{4} \bullet \frac{1}{8} = \frac{3}{32}$

68
Evaluate. $\frac{18}{19} \bullet \frac{5}{6}$

$\frac{3}{19} \bullet \frac{5}{1} = \frac{15}{19}$

69
To use cancellation with $\frac{-8}{5} \bullet \frac{7}{12}$ first find the HCF of 8 and 12.

$$\frac{-8}{5} \bullet \frac{7}{12} = \frac{\overset{-2}{\cancel{-8}}}{5} \bullet \frac{7}{\underset{3}{\cancel{12}}} = \frac{-2}{5} \bullet \frac{7}{3} = \frac{-14}{15}$$

Evaluate. $\frac{-7}{3} \bullet \frac{11}{14}$

$\frac{-1}{3} \bullet \frac{11}{2} = \frac{-11}{6}$

70
Evaluate. $\frac{-8}{3} \bullet \frac{-7}{16}$

$\frac{-1}{3} \bullet \frac{-7}{2} = \frac{7}{6}$

71
Evaluate. $\frac{6}{5} \bullet \frac{-4}{9}$

$\frac{-8}{15}$

72
Evaluate. $\frac{-9}{2} \bullet \frac{-5}{12}$

$\frac{15}{8}$

73
Evaluate. $\frac{4}{7} \bullet \frac{3}{16}$

$\frac{3}{28}$

74
To evaluate $\frac{5}{3} \bullet \frac{9}{10}$ two pairs of cancellations are used.

$$\frac{5}{3} \bullet \frac{9}{10} = \frac{\overset{1}{\cancel{5}}}{\underset{1}{\cancel{3}}} \bullet \frac{\overset{3}{\cancel{9}}}{\underset{2}{\cancel{10}}} = \frac{1}{1} \bullet \frac{3}{2} = \frac{3}{2}$$

Evaluate. $\frac{7}{5} \bullet \frac{25}{14}$

$\frac{1}{1} \bullet \frac{5}{2} = \frac{5}{2}$

75
Evaluate. $\frac{2}{15} \bullet \frac{25}{4}$

$\frac{1}{3} \bullet \frac{5}{2} = \frac{5}{6}$

76
Evaluate. $\frac{8}{9} \bullet \frac{3}{4}$

$\frac{2}{3} \bullet \frac{1}{1} = \frac{2}{3}$

77
Evaluate. $\frac{9}{20} \bullet \frac{10}{27}$

$\frac{1}{6}$

78
Evaluate. $\frac{4}{10} \bullet \frac{-5}{12}$

$\frac{-1}{6}$

79
The evaluation of $\frac{9}{7} \bullet \frac{14}{3}$ results in an integer.

$$\frac{9}{7} \bullet \frac{14}{3} = \frac{\overset{3}{\cancel{9}}}{\underset{1}{\cancel{7}}} \bullet \frac{\overset{2}{\cancel{14}}}{\underset{1}{\cancel{3}}} = \frac{3}{1} \bullet \frac{2}{1} = \frac{6}{1} = 6$$

Evaluate. $\frac{8}{3} \bullet \frac{9}{2}$

12

80
Evaluate. $\frac{-12}{5} \bullet \frac{25}{6}$

-10

81
The evaluation of $\frac{5}{2} \bullet \frac{2}{5}$ produces 1 as its result.

$$\frac{5}{2} \bullet \frac{2}{5} = \frac{\overset{1}{\cancel{5}}}{\underset{1}{\cancel{2}}} \bullet \frac{\overset{1}{\cancel{2}}}{\underset{1}{\cancel{5}}} = \frac{1}{1} \bullet \frac{1}{1} = \frac{1}{1} = 1$$

Evaluate. $\frac{5}{9} \bullet \frac{9}{5}$

1

82
Evaluate. $\frac{-7}{4} \bullet \frac{-4}{7}$

1

83
Evaluate. $\frac{-4}{5} \cdot \frac{-15}{2}$ — 6

84
Evaluate. $\frac{5}{12} \cdot \frac{16}{15}$ — $\frac{4}{9}$

85
Evaluate. $\frac{6}{15} \cdot \frac{25}{24}$ — $\frac{5}{12}$

86
Evaluate. $\frac{14}{3} \cdot \frac{18}{7}$ — 12

87
Evaluate. $\frac{-7}{8} \cdot \frac{16}{21}$ — $\frac{-2}{3}$

88
Evaluate. $\frac{9}{11} \cdot \frac{11}{9}$ — 1

89
Evaluate. $\frac{10}{7} \cdot \frac{-3}{5}$ — $\frac{-6}{7}$

90
Evaluate. $\frac{-10}{33} \cdot \frac{-11}{15}$ — $\frac{2}{9}$

FEEDBACK UNIT 4

This quiz reviews the preceding unit. Answers are at the back of the book.

Evaluate and write the answer in simplest form.

1. $\frac{8}{7} \cdot \frac{-3}{5}$
2. $\frac{2}{5} \cdot \frac{3}{7}$
3. $\frac{-5}{7} \cdot \frac{9}{8}$
4. $\frac{-6}{5} \cdot \frac{7}{6}$
5. $\frac{2}{5} \cdot \frac{15}{14}$
6. $\frac{-2}{5} \cdot \frac{-3}{11}$
7. $\frac{12}{5} \cdot \frac{-10}{9}$
8. $\frac{5}{7} \cdot \frac{7}{5}$
9. $\frac{-9}{4} \cdot \frac{-4}{9}$

UNIT 5: WRITING PAIRS OF RATIONAL NUMBERS WITH A COMMON DENOMINATOR

The following mathematical terms are crucial to an understanding of this unit.

Highest common factor, HCF
Changing names of rational numbers
Least common multiple, LCM

1

The rational number $\frac{2}{3}$ is equal to $\frac{8}{12}$. To change $\frac{2}{3}$ to $\frac{8}{12}$, the following steps are used:

$$\frac{2}{3} = \frac{2}{3} \bullet 1 = \frac{2}{3} \bullet \frac{4}{4} = \frac{8}{12}$$

To change $\frac{2}{3}$ to $\frac{8}{12}$, both the numerator and denominator of $\frac{2}{3}$ were multiplied by ______ .

4

2

To change $\frac{2}{3}$ to $\frac{14}{21}$, the desired denominator is 7 times the denominator of $\frac{2}{3}$.

$$\frac{2}{3} = \frac{2}{3} \bullet 1 = \frac{2}{3} \bullet \frac{7}{7} = \frac{14}{21}$$

To change $\frac{2}{3}$ to $\frac{14}{21}$, both the numerator and denominator of $\frac{2}{3}$ were multiplied by ______ .

7

3

To change $\frac{2}{3}$ to $\frac{?}{30}$, both the numerator and denominator of $\frac{2}{3}$ are multiplied by ______ .

10

4

$$\frac{2}{3} = \frac{2}{3} \bullet 1 = \frac{2}{3} \bullet \frac{10}{10} = \frac{20}{30}$$

To change $\frac{2}{3}$ to $\frac{?}{15}$, both the numerator and denominator of $\frac{2}{3}$ are multiplied by ______ .

5

5

Change $\frac{5}{6}$ to $\frac{?}{18}$, by multiplying both the numerator and the denominator by the same number.

$\frac{5}{6} \bullet \frac{3}{3} = \frac{15}{18}$

6

Change $\frac{5}{6}$ to $\frac{?}{42}$, by multiplying both the numerator and the denominator by the same number.

$\frac{5}{6} \bullet \frac{7}{7} = \frac{35}{42}$

7

Change $\frac{2}{5}$ to $\frac{?}{30}$, by multiplying both the numerator and the denominator by the same number.

$\frac{2}{5} \bullet \frac{6}{6} = \frac{12}{30}$

8

Change. $\frac{2}{5}$ to $\frac{?}{45}$

$\frac{2}{5} \bullet \frac{9}{9} = \frac{18}{45}$

9

Change. $\frac{2}{7}$ to $\frac{?}{35}$

$\frac{2}{7} \bullet \frac{5}{5} = \frac{10}{35}$

10

Change. $\frac{-5}{7}$ to $\frac{?}{21}$

$\frac{-5}{7} \bullet \frac{3}{3} = \frac{-15}{21}$

11

To **change** $\frac{3}{-4}$ to $\frac{?}{20}$, first write $\frac{3}{-4}$ with a **positive denominator**.

$$\frac{3}{-4} = \frac{-3}{4} \bullet \frac{5}{5} = \frac{-15}{20}$$

Change. $\frac{5}{-6}$ to $\frac{?}{24}$

$\frac{-5}{6} \bullet \frac{4}{4} = \frac{-20}{24}$

12

Change. $\frac{8}{-9}$ to $\frac{?}{36}$

$\frac{-8}{9} \bullet \frac{4}{4} = \frac{-32}{36}$

13

To change $\frac{-1}{-6}$ to $\frac{?}{24}$, first write $\frac{-1}{-6}$ with a positive denominator.

$$\frac{-1}{-6} = \frac{1}{6} \bullet \frac{4}{4} = \frac{4}{24}$$

Change. $\frac{-8}{-11}$ to $\frac{?}{33}$

$\frac{8}{11} \bullet \frac{3}{3} = \frac{24}{33}$

14

Change. $\frac{3}{4}$ to $\frac{?}{8}$

$\frac{6}{8}$

15

Change. $\frac{-9}{10}$ to $\frac{?}{30}$

$\frac{-27}{30}$

16
Change. $\frac{9}{4}$ to $\frac{?}{20}$ — $\frac{45}{20}$

17
Change. $\frac{2}{-9}$ to $\frac{?}{63}$ — $\frac{-14}{63}$

18
Change. $\frac{5}{8}$ to $\frac{?}{24}$ — $\frac{15}{24}$

19
Change. $\frac{7}{10}$ to $\frac{?}{40}$ — $\frac{28}{40}$

20
Change. $\frac{7}{5}$ to $\frac{?}{30}$ — $\frac{42}{30}$

21
Change. $\frac{-2}{3}$ to $\frac{?}{12}$ — $\frac{-8}{12}$

22
The factors of 12 are: 1, 2, 3, 4, 6, 12.
The factors of 18 are: ______________. — 1, 2, 3, 6, 9, 18

23
6 is the **highest common factor, HCF**, of 12 and 18 because 6 is the largest number that is a factor of 12 and also a factor of ______. — 18

24
The factors of 10 are: 1, 2, 5, 10.
The factors of 25 are: ___________. — 1, 5, 25

25
The HCF of 10 and 25 is ______ . — 5

26
The HCF of 10 and 12 is ______. — 2

27
The HCF of 4 and 8 is ______. — 4

28
The HCF of 12 and 30 is ______. — 6

29
The HCF of 35 and 14 is ______. — 7

30
To find the **least common multiple, LCM,** of 12 and 18, the following steps are used.

1. Find the HCF of 12 and 18.
2. Divide the product of 12 and 18 (12 • 18) by the HCF.

What is the HCF of 12 and 18?

6

31
The LCM of 12 and 18 is found by:

1. Finding the HCF of 12 and 18 which is 6.
2. Dividing the product of 12 and 18 by the HCF.

The division required in step 2 is often simplified by the cancellation process.

$$\frac{12 \cdot 18}{6} = \frac{\overset{2}{\cancel{12}} \cdot 18}{\underset{1}{\cancel{6}}} = \frac{2 \cdot 18}{1} = 2 \cdot 18 = 36$$

The LCM of 12 and 18 is ______.

36

32
Shown below are two cancellation problems.

$$\frac{\overset{2}{\cancel{12}} \cdot 18}{\underset{1}{\cancel{6}}} = 2 \cdot 18 = 36 \text{ and } \frac{12 \cdot \overset{3}{\cancel{18}}}{\underset{1}{\cancel{6}}} = 12 \cdot 3 = 36$$

Does the cancellation method effect the final result?

No, both are 36

33
To find the least common multiple (LCM) of 10 and 25:

1. Find the HCF of 10 and 25.
2. Divide the product of 10 and 25 (10 • 25) by the HCF.

What is the HCF of 10 and 25?

5

34
To find the LCM of 10 and 25, divide the product of the numbers by their HCF.

$$\frac{\overset{2}{\cancel{10}} \cdot 25}{\underset{1}{\cancel{5}}} = 2 \cdot 25 = 50$$

What is the least common multiple (LCM) of 10 and 25?

50

35
Find the least common multiple (LCM) of 15 and 25 by dividing their product by their HCF.

$$\frac{15 \cdot 25}{5} = 75$$

36
Find the least common multiple (LCM) of 12 and 20 by dividing their product by their HCF.

$\frac{12 \cdot 20}{4} = 60$

37
To write $\frac{5}{12}$ and $\frac{11}{18}$ with the same denominators:
1. Find the LCM of 12 and 18.
2. Write each fraction with the LCM as the denominator.

What is the LCM of 12 and 18?

$\frac{12 \cdot 18}{6} = 36$

38
To write $\frac{5}{12}$ and $\frac{11}{18}$ with the same denominators:
1. Find the LCM of 12 and 18 which is 36.
2. Write each fraction with 36 as its denominator.

Change. $\frac{5}{12}$ to $\frac{?}{36}$

$\frac{5}{12} \cdot \frac{3}{3} = \frac{15}{36}$

39
The LCM of 12 and 18 is 36.
$\frac{5}{12} = \frac{15}{36}$ $\frac{11}{18} =$ _______

$\frac{22}{36}$

40
To write $\frac{7}{10}$ and $\frac{3}{25}$ with equal denominators:
1. Find the LCM of 10 and 25.
2. Write each fraction with the LCM as its denominator.

What is the LCM of 10 and 25?

$\frac{10 \cdot 25}{5} = 50$

40
To write $\frac{7}{10}$ and $\frac{3}{25}$ with a common denominator:
1. Find the LCM of 10 and 25 which is 50.
2. Write each fraction with 50 as its denominator.

$\frac{7}{10} = \frac{35}{50}$ $\frac{3}{25} =$ _______

$\frac{6}{50}$

41
The first step in writing $\frac{3}{4}$ and $\frac{5}{6}$ with a common denominator is to find the least common multiple of 4 and 6. What is the LCM of 4 and 6?

$\frac{4 \cdot 6}{2} = 12$

42
The common denominator for $\frac{3}{4}$ and $\frac{5}{6}$ is 12.
$\frac{3}{4} = \frac{\quad}{12}$ $\frac{5}{6} = \frac{\quad}{12}$

$\frac{9}{12}, \frac{10}{12}$

43
As the first step to writing $\frac{5}{8}$ and $\frac{3}{4}$ with a common denominator, what is the LCM of 8 and 4.

$\frac{8 \cdot 4}{4} = 8$

44
The common denominator for $\frac{5}{8}$ and $\frac{3}{4}$ is 8.

$\frac{5}{8} = \frac{\quad}{8}$ $\frac{3}{4} = \frac{\quad}{8}$

$\frac{5}{8}, \frac{6}{8}$

45
As the first step to writing $\frac{2}{3}$ and $\frac{1}{4}$ with a common denominator, what is the LCM of 3 and 4.

$\frac{3 \cdot 4}{1} = 12$

46
The common denominator for $\frac{2}{3}$ and $\frac{1}{4}$ is 12.

$\frac{2}{3} = \frac{\quad}{12}$ $\frac{1}{4} = \frac{\quad}{12}$

$\frac{8}{12}, \frac{3}{12}$

47
As the first step to writing $\frac{5}{8}$ and $\frac{-3}{10}$ with a common denominator, what is the LCM of 8 and 10.

$\frac{8 \cdot 10}{2} = 40$

48
The common denominator for $\frac{5}{8}$ and $\frac{-3}{10}$ is 40.

$\frac{5}{8} = \frac{\quad}{40}$ $\frac{-3}{10} = \frac{\quad}{40}$

$\frac{25}{40}, \frac{-12}{40}$

49
What is the common denominator for $\frac{-5}{9}$ and $\frac{-1}{6}$?

18

50
Write $\frac{-5}{9}$ and $\frac{-1}{6}$ with a common denominator.

$\frac{-10}{18}, \frac{-3}{18}$

51
What is the common denominator for $\frac{3}{8}$ and $\frac{5}{12}$?

24

52
Write $\frac{3}{8}$ and $\frac{5}{12}$ with a common denominator.

$\frac{9}{24}, \frac{10}{24}$

53
What is the common denominator for $\frac{5}{6}$ and $\frac{2}{7}$?

42

54
Write $\frac{5}{6}$ and $\frac{2}{7}$ with a common denominator.

$\frac{35}{42}, \frac{12}{42}$

55
What is the common denominator for $\frac{7}{10}$ and $\frac{9}{4}$?

20

56
Write $\frac{7}{10}$ and $\frac{9}{4}$ with a common denominator.

$\frac{14}{20}, \frac{45}{20}$

57
Write $\frac{5}{8}$ and $\frac{3}{14}$ with a common denominator.

$\frac{35}{56}, \frac{12}{56}$

58
Write $\frac{-3}{5}$ and $\frac{7}{9}$ with a common denominator.

$\frac{-27}{45}, \frac{35}{45}$

59
Write $\frac{7}{8}$ and $\frac{-5}{12}$ with a common denominator.

$\frac{21}{24}, \frac{-10}{24}$

60
Write $\frac{7}{10}$ and $\frac{4}{15}$ with a common denominator.

$\frac{21}{30}, \frac{8}{30}$

61
Write $\frac{4}{5}$ and $\frac{-6}{11}$ with a common denominator.

$\frac{44}{55}, \frac{-30}{55}$

62
Write $\frac{3}{4}$ and $\frac{-5}{18}$ with a common denominator.

$\frac{27}{36}, \frac{-10}{36}$

FEEDBACK UNIT 5

This quiz reviews the preceding unit. Answers are at the back of the book.

Write each pair of fractions with a common denominator.

1. $\frac{1}{20}, \frac{3}{25}$
2. $\frac{7}{8}, \frac{3}{10}$
3. $\frac{3}{5}, \frac{9}{4}$
4. $\frac{5}{8}, \frac{-7}{12}$
5. $\frac{-3}{10}, \frac{4}{7}$
6. $\frac{-3}{8}, \frac{17}{24}$

UNIT 6: ADDING RATIONAL NUMBERS

1
To add rational numbers with a common denominator, add the numerators

$$\frac{3}{5} + \frac{1}{5} = \frac{3+1}{5} = \frac{4}{5}$$

Evaluate $\frac{5}{7} + \frac{1}{7}$ by adding the numerators.

$\frac{6}{7}$

2
To add rational numbers with a common denominator, add the numerators and keep the same denominator.

$$\frac{7}{10} + \frac{6}{10} = \frac{7+6}{10} = \frac{13}{10}$$

Evaluate $\frac{5}{9} + \frac{8}{9}$ by adding the numerators.

$\frac{13}{9}$

3
To add $\frac{5}{8}$ and $\frac{3}{8}$, the numerators are added.

$$\frac{5}{8} + \frac{3}{8} = \frac{5+3}{8} = \frac{8}{8}$$

Can $\frac{8}{8}$ be simplified?

Yes, it is 1.

4
Add. $\frac{3}{10}$ and $\frac{7}{10}$

$\frac{10}{10} = 1$

5
To add $\frac{9}{11}$ and $\frac{7}{11}$, the numerators are added.

$$\frac{9}{11} + \frac{7}{11} = \frac{9+7}{11} = \frac{16}{11}$$

Add. $\frac{3}{7}$ and $\frac{5}{7}$

$\frac{8}{7}$

6
To add $\frac{4}{5}$ and $\frac{3}{7}$, the first step is to find a common denominator. What is the LCM of 5 and 7?

35

7
To add (evaluate) $\frac{4}{5} + \frac{3}{7}$, write each number with 35 as its denominator.

$$\frac{4}{5} = \frac{28}{35} \quad \text{and} \quad \frac{3}{7} = \frac{\quad}{35}$$

$\frac{15}{35}$

8

To evaluate $\frac{4}{5} + \frac{3}{7}$, begin by writing the numbers with a common denominator.

$$\begin{array}{ccc} \frac{4}{5} & + & \frac{3}{7} \\ \downarrow & & \downarrow \\ \frac{28}{35} & + & \frac{15}{35} \end{array}$$

Complete the evaluation of $\frac{4}{5} + \frac{3}{7}$.

$\frac{28}{35} + \frac{15}{35} = \frac{43}{35}$

9

To evaluate $\frac{3}{8} + \frac{5}{12}$, a common multiple for 8 and 12 must first be found. What is the LCM of 8 and 12?

24

10

Evaluate $\frac{3}{8} + \frac{5}{12}$, by first writing each number with a common denominator.

$$\begin{array}{ccc} \frac{3}{8} & + & \frac{5}{12} \\ \downarrow & & \downarrow \\ \frac{9}{24} & + & \frac{\quad}{24} \end{array}$$

$\frac{10}{24}$

11

To evaluate $\frac{3}{8} + \frac{5}{12}$, the following steps are used:

$$\begin{array}{ccc} \frac{3}{8} & + & \frac{5}{12} \\ \downarrow & & \downarrow \\ \frac{9}{24} & + & \frac{10}{24} \end{array}$$

Complete the evaluation of $\frac{3}{8} + \frac{5}{12}$.

$\frac{9}{24} + \frac{10}{24} = \frac{19}{24}$

12

Complete the evaluation of $\frac{3}{5} + \frac{7}{11}$.

$$\frac{3}{5} + \frac{7}{11} = \frac{33}{55} + \frac{35}{55} = \_\_\_\_\_\_\_\_$$

$\frac{68}{55}$

13

The following example depends upon the fact that 25 + (-4) equals 21.

$$\frac{5}{6} + \frac{-2}{15} = \frac{25}{30} + \frac{-4}{30} = \frac{21}{30} = \frac{7}{10}$$

Evaluate. $\frac{7}{15} + \frac{-1}{6}$

$\frac{9}{30} = \frac{3}{10}$

14

Complete the evaluation of $\frac{1}{4} + \frac{5}{12}$ started below.

$\frac{1}{4} + \frac{5}{12} = \frac{3}{12} + \frac{5}{12} =$ __________

$\frac{8}{12} = \frac{2}{3}$

15

Complete the evaluation of $\frac{-5}{8} + \frac{2}{9}$ started below.

$\frac{-5}{8} + \frac{2}{9} = \frac{-45}{72} + \frac{16}{72} =$ __________

$\frac{-29}{72}$

16

Evaluate. $\frac{3}{7} + \frac{2}{5}$

$\frac{15}{35} + \frac{14}{35} = \frac{29}{35}$

17

Evaluate. $\frac{3}{8} + \frac{5}{12}$

$\frac{9}{24} + \frac{10}{24} = \frac{19}{24}$

18

Evaluate. $\frac{4}{3} + \frac{1}{7}$

$\frac{28}{21} + \frac{3}{21} = \frac{31}{21}$

19

Evaluate. $\frac{1}{2} + \frac{7}{10}$

$\frac{5}{10} + \frac{7}{10} = \frac{12}{10} = \frac{6}{5}$

20

Evaluate. $\frac{4}{9} + \frac{2}{5}$

$\frac{38}{45}$

21

Evaluate. $\frac{9}{17} + \frac{-2}{17}$

$\frac{7}{17}$

22

Evaluate $\frac{3}{4} + 5$ by first writing 5 as $\frac{5}{1}$.

$\frac{23}{4}$

23

Evaluate. $\frac{2}{7} + 3$

$\frac{23}{7}$

24

Evaluate. $\frac{1}{8} + \frac{2}{3}$

$\frac{19}{24}$

25

Evaluate. $\frac{5}{6} + \frac{1}{5}$

$\frac{31}{30}$

26

Evaluate. $\frac{3}{14} + \frac{1}{4}$

$\frac{13}{28}$

27

Evaluate. $\frac{3}{5} + 2$

$\frac{13}{5}$

28
Evaluate. $\frac{2}{5} + \frac{4}{7}$

$\frac{34}{35}$

29
Evaluate. $\frac{1}{6} + \frac{1}{3}$

$\frac{3}{6} = \frac{1}{2}$

To add two rational numbers:

1. Write both rational numbers with the same denominator.
2. Add the numerators, and place the sum over the denominator.

$$\frac{5}{8} + \frac{3}{10} = \frac{25}{40} + \frac{12}{40} = \frac{37}{40}$$

30
Evaluate. $\frac{2}{9} + \frac{-2}{5}$

$\frac{-8}{45}$

31
Evaluate. $\frac{-5}{3} + \frac{-3}{4}$

$\frac{-29}{12}$

32
Evaluate. $\frac{-2}{3} + \frac{-4}{5}$

$\frac{-22}{15}$

33
Evaluate $\frac{-2}{7} + \frac{4}{-3}$ by first writing $\frac{4}{-3}$ as $\frac{-4}{3}$.

$\frac{-34}{21}$

34
Evaluate $\frac{2}{-3} + \frac{-5}{8}$ by first writing $\frac{2}{-3}$ as $\frac{-2}{3}$.

$\frac{-31}{24}$

35
Evaluate. $\frac{1}{-2} + \frac{5}{8}$

$\frac{1}{8}$

36
Evaluate. $\frac{3}{-4} + \frac{2}{-5}$

$\frac{-23}{20}$

37
Evaluate. $\frac{2}{-7} + \frac{5}{-3}$

$\frac{-41}{21}$

38
Evaluate. $\frac{3}{-8} + \frac{1}{-6}$

$\frac{-13}{24}$

39
Evaluate $\frac{-3}{-2} + \frac{5}{3}$ by first writing $\frac{-3}{-2}$ as $\frac{3}{2}$.

$\frac{19}{6}$

40
Evaluate. $\frac{-7}{5} + \frac{-3}{-4}$ $\frac{-13}{20}$

41
Evaluate $\frac{3}{7} + 0$ by first writing 0 as $\frac{0}{1}$. $\frac{3}{7}$

42
Evaluate. $\frac{-6}{5} + 0$ $\frac{-6}{5}$

43
Evaluate. $\frac{3}{10} + \frac{-5}{6}$ $\frac{-16}{30} = \frac{-8}{15}$

44
Evaluate. $\frac{5}{7} + \frac{-5}{7}$ $\frac{0}{7} = 0$

45
Evaluate. $\frac{-4}{9} + \frac{4}{9}$ 0

46
Whenever two numbers have a sum of zero, the numbers are **opposites**. -5 + 5 = 0. Therefore, -5 and 5 are ______. opposites

47
Are $\frac{3}{8}$ and $\frac{-3}{8}$ opposites? Yes, $\frac{3}{8} + \frac{-3}{8} = 0$

48
Are $\frac{-2}{11}$ and $\frac{2}{11}$ opposites? Yes

49
Are $\frac{2}{9}$ and $\frac{-9}{2}$ opposites? No, $\frac{2}{9} + \frac{-9}{2} = \frac{-77}{18}$

50
Are $\frac{-17}{25}$ and $\frac{17}{25}$ opposites? Yes

51
To find the opposite of any rational number:

1. Change the numerator to its opposite.
2. Do not change the denominator.

The opposite of $\frac{7}{16}$ is ______. $\frac{-7}{16}$

52
The opposite of $\frac{-13}{5}$ is ______. $\frac{13}{5}$

53
The opposite of $\frac{-4}{17}$ is ______. $\frac{4}{17}$

54

Recall that 5 – 11 means 5 + (-11) because the minus sign indicates that the opposite of the second number is to be added to the first. The same use of the minus sign applies to the rational numbers. $\frac{2}{3} - \frac{7}{12}$ means $\frac{2}{3}$ + ________.

$\frac{-7}{12}$

55

$\frac{2}{7} - \frac{1}{3}$ means $\frac{2}{7} + \frac{-1}{3}$. The minus sign between any two rational numbers indicates that the opposite of the second is to be added to the first.

$\frac{3}{4} - \frac{2}{5}$ means ______.

$\frac{3}{4} + \frac{-2}{5}$

56

$\frac{-3}{5} - \frac{1}{4}$ means $\frac{-3}{5} + \frac{-1}{4}$

$\frac{-6}{7} - \frac{2}{3}$ means __________

$\frac{-6}{7} + \frac{-2}{3}$

57

$\frac{3}{8} - \frac{2}{5}$ means __________

$\frac{3}{8} + \frac{-2}{5}$

58

$\frac{-8}{3} - \frac{4}{13}$ means __________

$\frac{-8}{3} + \frac{-4}{13}$

59

Recall that 6 – (-5) means 6 + 5 because the opposite of -5 is 5. Similarly, $\frac{3}{4} - \frac{-5}{8}$ means ________.

$\frac{3}{4} + \frac{5}{8}$

60

$\frac{-7}{8} - \frac{-3}{5}$ means $\frac{-7}{8} + \frac{3}{5}$

$\frac{-8}{11} - \frac{-4}{7}$ means __________

$\frac{-8}{11} + \frac{4}{7}$

61

Evaluate. $\frac{5}{7} - \frac{8}{7}$

$\frac{5}{7} + \frac{-8}{7} = \frac{-3}{7}$

62

Evaluate. $\frac{9}{11} - \frac{-7}{11}$

$\frac{9}{11} + \frac{7}{11} = \frac{16}{11}$

63

Evaluate. $\frac{2}{13} - \frac{-5}{13}$

$\frac{2}{13} + \frac{5}{13} = \frac{7}{13}$

64

To evaluate $\frac{3}{4} - \frac{1}{3}$, the following steps are used:

$$\frac{3}{4} - \frac{1}{3} = \frac{3}{4} + \frac{-1}{3} = \frac{9}{12} + \frac{-4}{12} = \frac{5}{12}$$

Evaluate. $\frac{3}{10} - \frac{1}{7}$

$\frac{21}{70} + \frac{-10}{70} = \frac{11}{70}$

65

Complete the evaluation: $\frac{4}{3} - \frac{2}{5} = \frac{4}{3} + \frac{-2}{5} =$ ______

$\frac{20}{15} + \frac{-6}{15} = \frac{14}{15}$

66

Evaluate. $\frac{5}{6} - \frac{2}{5}$

$\frac{13}{30}$

67

Evaluate. $\frac{6}{7} - \frac{1}{3}$

$\frac{11}{21}$

68

To evaluate $\frac{4}{5} - \frac{-1}{3}$, the following steps are used:

$$\frac{4}{5} - \frac{-1}{3} = \frac{4}{5} + \frac{1}{3} = \frac{12}{15} + \frac{5}{15} = \frac{17}{15}$$

Evaluate. $\frac{2}{7} - \frac{-1}{4}$

$\frac{8}{28} + \frac{7}{28} = \frac{15}{28}$

69

Complete the evaluation: $\frac{7}{8} - \frac{-1}{5} = \frac{7}{8} + \frac{1}{5} =$ ______

$\frac{35}{40} + \frac{8}{40} = \frac{43}{40}$

70

Evaluate. $\frac{4}{7} - \frac{-2}{5}$

$\frac{34}{35}$

71

Evaluate. $\frac{3}{4} - \frac{-1}{8}$

$\frac{7}{8}$

72

Evaluate. $\frac{-3}{5} - \frac{1}{6}$

$\frac{-23}{30}$

73

Evaluate. $\frac{-7}{8} - \frac{-2}{5}$

$\frac{-19}{40}$

74

Evaluate. $\frac{-5}{6} - \frac{7}{8}$

$\frac{-41}{24}$

75

Evaluate. $\frac{5}{11} + \frac{-4}{3}$

$\frac{-29}{33}$

76

Evaluate. $\frac{3}{10} - \frac{7}{15}$

$\frac{-5}{30} = \frac{-1}{6}$

77
Evaluate. $\frac{5}{8} - \frac{-2}{3}$

$\frac{31}{24}$

78
Evaluate. $\frac{-5}{12} - \frac{11}{18}$

$\frac{-37}{36}$

FEEDBACK UNIT 6

This quiz reviews the preceding unit. Answers are at the back of the book.

1. Evaluate. $\frac{5}{4} + \frac{1}{7}$
2. Evaluate. $\frac{-1}{2} + \frac{-3}{4}$
3. Evaluate. $\frac{5}{6} + \frac{1}{9}$
4. Evaluate. $5 + \frac{-1}{7}$
5. Evaluate. $\frac{5}{9} + 0$
6. The opposite of $\frac{5}{7}$ is ______.
7. The opposite of $\frac{-8}{3}$ is ______.
8. $\frac{9}{10} - \frac{3}{25} =$ ______.
9. $\frac{3}{8} - \frac{-1}{5} =$ ______.

UNIT 7: RECIPROCALS AND DIVISION OF RATIONAL NUMBERS

The following mathematical terms are crucial to an understanding of this unit.

Reciprocals
Complex fractions
Division of rationals

1
Recall that the multiplication of rationals requires separately multiplying numerators and denominators.

Evaluate. $\frac{7}{9} \bullet \frac{5}{8}$

$\frac{7}{9} \bullet \frac{5}{8} = \frac{7 \bullet 5}{9 \bullet 8} = \frac{35}{72}$

2
Whenever possible, simplify the multiplication of rationals by cancelling. Evaluate $\frac{4}{7} \bullet \frac{3}{4}$.

$\frac{1}{7} \bullet \frac{3}{1} = \frac{3}{7}$

3
For the rational number $\frac{3}{10}$, if the numerator and denominator are interchanged the rational number $\frac{10}{3}$ is obtained. $\frac{3}{10} \bullet \frac{10}{3} =$ ______.

1

4
For the rational number $\frac{7}{8}$, what number is obtained by interchanging the numerator and denominator?

$\frac{8}{7}$

5
Evaluate. $\frac{7}{8} \bullet \frac{8}{7}$

1

6
What rational number is obtained if the numerator and denominator of $\frac{5}{9}$ are interchanged?

$\frac{9}{5}$

7
Evaluate. $\frac{5}{9} \bullet \frac{9}{5}$

1

8
For the open sentence $\frac{2}{7}x = 1$, if x is replaced by $\frac{7}{2}$ the statement $\frac{2}{7} \bullet \frac{7}{2} = 1$ is obtained. Is the statement $\frac{2}{7} \bullet \frac{7}{2} = 1$ true?

Yes

9
For the open sentence $\frac{7}{6}x = 1$, if x is replaced by $\frac{6}{7}$ will the statement $\frac{7}{6} \bullet \frac{6}{7} = 1$ be true?

Yes

10
What rational number can replace x to make the following equation a true statement? $\frac{3}{4}x = 1$

$\frac{4}{3}$

11
What rational number can replace z to make the equation $\frac{5}{8}z = 1$ a true statement?

$\frac{8}{5}$

12
For $\frac{-2}{5}$, if the numerator and denominator are interchanged, $\frac{5}{-2}$ is obtained. $\frac{5}{-2}$ can be simplified to $\frac{-5}{2}$. Evaluate. $\frac{-2}{5} \cdot \frac{-5}{2}$

1

13
For $\frac{-8}{11}$, what rational number is obtained if the numerator and denominator are interchanged?

$\frac{11}{-8} = \frac{-11}{8}$

14
Evaluate. $\frac{-8}{11} \cdot \frac{-11}{8}$

1

15
Evaluate. $\frac{-9}{4} \cdot \frac{-4}{9}$

1

16
What rational number can replace y in the equation $\frac{-5}{12}y = 1$ to make a true statement?

$\frac{-12}{5}$

17
What rational number can replace w in the equation $\frac{-1}{2}w = 1$ to make a true statement?

$\frac{-2}{1} = -2$

18
Whenever two rational numbers have a product of 1, they are **reciprocals**. $\frac{2}{3}$ is the __________ of $\frac{3}{2}$ because $\frac{2}{3} \cdot \frac{3}{2} = 1$.

reciprocal

19
$\frac{3}{4}$ is the reciprocal of $\frac{4}{3}$. What is the reciprocal of $\frac{5}{13}$?

$\frac{13}{5}$, because $\frac{5}{13} \cdot \frac{13}{5} = 1$

20
The reciprocal of a rational number is obtained by interchanging the numerator and denominator. What is the reciprocal of $\frac{1}{5}$?

$\frac{5}{1} = 5$

21
What is the reciprocal of $\frac{3}{8}$?

$\frac{8}{3}$

22
What is the reciprocal of 7?

$\frac{1}{7}$, because $7 \cdot \frac{1}{7} = 1$

23

The reciprocal of $\frac{-3}{4}$ is $\frac{-4}{3}$. $\frac{-3}{4} \bullet \frac{-4}{3} = 1$

What is the reciprocal of $\frac{-3}{5}$?

$\frac{-5}{3}$

24

What is the reciprocal of $\frac{-5}{17}$?

$\frac{-17}{5}$

25

What is the reciprocal of -3?

$\frac{-1}{3}$, because $-3 \bullet \frac{-1}{3} = 1$

26

What is the reciprocal of $\frac{9}{2}$?

$\frac{2}{9}$

27

What is the reciprocal of 8?

$\frac{1}{8}$

28

Every rational number except zero has a reciprocal. Why is there no reciprocal for zero?

0 times any number is 0, not 1.

29

$\frac{5}{7}$ and $\frac{7}{5}$ are reciprocals because $\frac{5}{7} \bullet \frac{7}{5} = 1$.

The product of any rational number and its reciprocal is ______.

1

30

Does every rational number except zero have a reciprocal?

Yes

31

Does $\frac{0}{1}$ have a reciprocal?

No, because $\frac{0}{1} = 0$

32

Every rational number except ______ has a reciprocal.

zero

33

The product of any rational number and its reciprocal is ______.

1

34

The reciprocal of a rational number can be found by interchanging the numerator and denominator. The reciprocal of $\frac{7}{10}$ is ______.

$\frac{10}{7}$

35

The reciprocal of $\frac{-4}{11}$ is ______.

$\frac{-11}{4}$

36

The **division problem** $2 \div 9$ can be written as $2 \bullet \frac{1}{9}$ where $\frac{1}{9}$ is the reciprocal of 9. Write $3 \div 8$ as a multiplication problem using the reciprocal of 8.

$3 \bullet \frac{1}{8}$

37

The division problem $-3 \div 5$ can be written as $-3 \bullet \frac{1}{5}$. Write $-7 \div 5$ as a multiplication problem.

$-7 \bullet \frac{1}{5}$

38

Every division problem can be written as a fraction or as a multiplication problem.

Division problem	$9 \div 4$
Fraction	$\frac{9}{4}$
Multiplication problem	$9 \bullet \frac{1}{4}$

Write $-5 \div 7$ as both a fraction and a multiplication problem.

$\frac{-5}{7}$, $-5 \bullet \frac{1}{7}$

39

The division of $\frac{3}{4}$ by 6 can be written in three ways:

Division problem	$\frac{3}{4} \div 6$
Fraction	$\frac{\frac{3}{4}}{6}$
Multiplication problem	$\frac{3}{4} \bullet \frac{1}{6}$

Write $\frac{2}{5} \div 3$ as both a fraction and a multiplication problem.

$\frac{\frac{2}{5}}{3}$, $\frac{2}{5} \bullet \frac{1}{3}$

40

The division problem $\frac{2}{5} \div \frac{3}{4}$ can be written as a **complex fraction** or as a multiplication problem.

Division problem	$\frac{2}{5} \div \frac{3}{4}$
Complex Fraction	$\dfrac{\frac{2}{5}}{\frac{3}{4}}$
Multiplication problem	$\frac{2}{5} \bullet \frac{4}{3}$

Write $\frac{-5}{8} \div \frac{2}{7}$ as both a complex fraction and a multiplication problem.

$\dfrac{\frac{-5}{8}}{\frac{2}{7}}$, $\frac{-5}{8} \bullet \frac{7}{2}$

41

The complex fraction $\dfrac{\frac{3}{4}}{\frac{5}{8}}$ is $\frac{3}{4} \div \frac{5}{8}$ or $\frac{3}{4} \bullet \frac{8}{5}$

Write $\dfrac{\frac{2}{3}}{\frac{4}{7}}$ as both a division and a multiplication problem.

$\frac{2}{3} \div \frac{4}{7}$, $\frac{2}{3} \bullet \frac{7}{4}$

42

The complex fraction $\dfrac{-3}{\frac{4}{5}}$ is $-3 \div \frac{4}{5}$ or $-3 \bullet \frac{5}{4}$

Write $\dfrac{4}{\frac{-5}{9}}$ as both a division and a multiplication problem.

$4 \div \frac{-5}{9}$, $4 \bullet \frac{-9}{5}$

43

The complex fraction $\dfrac{\frac{2}{3}}{-6}$ is $\frac{2}{3} \div -6$ or $\frac{2}{3} \bullet \frac{-1}{6}$.

Write $\dfrac{\frac{6}{5}}{-2}$ as both a division and a multiplication problem.

$\frac{6}{5} \div -2$, $\frac{6}{5} \bullet \frac{-1}{2}$

44

To evaluate a complex fraction, the following steps are used:

$$\dfrac{\frac{5}{8}}{\frac{2}{3}} = \frac{5}{8} \div \frac{2}{3} = \frac{5}{8} \bullet \frac{3}{2} = \frac{15}{16}$$

In the multiplication, why use $\frac{3}{2}$?

It is the reciprocal of $\frac{2}{3}$

45

Every complex fraction can be evaluated as a multiplication problem.

$$\frac{2}{\frac{-5}{6}} = 2 \bullet \frac{-6}{5} = \frac{-12}{5}$$

In the multiplication, why use $\frac{-6}{5}$?

It is the reciprocal of $\frac{-5}{6}$

46

Complex fractions are evaluated as multiplication problems.

$$\frac{\frac{5}{11}}{\frac{-2}{9}} = \frac{5}{11} \bullet \frac{-9}{2} = \frac{-45}{22}$$

Evaluate. $\frac{\frac{-2}{5}}{\frac{7}{8}}$

$\frac{-2}{5} \bullet \frac{8}{7} = \frac{-16}{35}$

47

Complex fractions (divisions) are evaluated as multiplication problems.

$$\frac{\frac{-5}{9}}{\frac{-3}{5}} = \frac{-5}{9} \bullet \frac{-5}{3} = \frac{25}{27}$$

Evaluate. $\frac{\frac{-8}{7}}{\frac{5}{3}}$

$\frac{-8}{7} \bullet \frac{3}{5} = \frac{-24}{35}$

48

Evaluate and simplify. $\frac{4}{\frac{2}{5}}$

$4 \bullet \frac{5}{2} = 2 \bullet 5 = 10$

49

Evaluate and simplify. $\frac{\frac{5}{6}}{\frac{-2}{9}}$

$\frac{5}{6} \bullet \frac{-9}{2} = \frac{5}{2} \bullet \frac{-3}{2} = \frac{-15}{4}$

50

Evaluate and simplify. $\frac{\frac{3}{4}}{7}$

$\frac{3}{4} \bullet \frac{1}{7} = \frac{3}{28}$

51

Evaluate and simplify. $\frac{\frac{4}{5}}{-2}$

$\frac{4}{5} \bullet \frac{-1}{2} = \frac{2}{5} \bullet -1 = \frac{-2}{5}$

52

Evaluate and simplify. $\frac{8}{\frac{1}{5}}$

$8 \cdot 5 = 40$

53

Evaluate. $\frac{-3}{\frac{4}{7}}$

$\frac{-21}{4}$

54

Evaluate. $\frac{\frac{8}{3}}{\frac{-3}{5}}$

$\frac{-40}{9}$

55

Evaluate. $\frac{\frac{-7}{8}}{\frac{-3}{4}}$

$\frac{7}{6}$

56

Evaluate. $\frac{\frac{-3}{7}}{\frac{11}{14}}$

$\frac{-6}{11}$

57

Evaluate. $\frac{\frac{6}{5}}{\frac{4}{3}}$

$\frac{9}{10}$

58

Evaluate. $\frac{\frac{-3}{8}}{\frac{-1}{4}}$

$\frac{3}{2}$

59

Evaluate. $\frac{\frac{2}{5}}{\frac{-1}{8}}$

$\frac{-16}{5}$

60

Evaluate. $\frac{\frac{-9}{2}}{\frac{-5}{2}}$

$\frac{9}{5}$

61

Evaluate. $\dfrac{\frac{-3}{4}}{\frac{-5}{12}}$

$\frac{9}{5}$

62

Evaluate. $\dfrac{8}{\frac{-2}{3}}$

-12

63

Evaluate. $\dfrac{\frac{8}{5}}{\frac{-1}{5}}$

-8

64

Evaluate. $\dfrac{\frac{10}{7}}{-4}$

$\frac{-5}{14}$

65

Evaluate. $\dfrac{\frac{-5}{7}}{\frac{7}{5}}$

$\frac{-25}{49}$

66

Evaluate. $\dfrac{\frac{-19}{24}}{\frac{1}{8}}$

$\frac{-19}{3}$

67

Evaluate. $\dfrac{-7}{\frac{2}{3}}$

$\frac{-21}{2}$

68

Evaluate. $\dfrac{\frac{-6}{7}}{\frac{6}{7}}$

-1

69

Evaluate. $\dfrac{-42}{\frac{-7}{3}}$

18

FEEDBACK UNIT 7

This quiz reviews the preceding unit. Answers are at the back of the book.

1. Find the reciprocals of each of the following.

   a. $\frac{2}{7}$ b. 4 c. $\frac{-9}{4}$ d. -7

Evaluate and simplify each of the following.

2. $\dfrac{\frac{6}{7}}{5}$ 3. $\dfrac{\frac{4}{5}}{-3}$ 4. $\dfrac{\frac{-2}{7}}{-2}$ 5. $\dfrac{8}{\frac{3}{5}}$

6. $\dfrac{-5}{\frac{4}{3}}$ 7. $\dfrac{\frac{3}{4}}{\frac{2}{5}}$

UNIT 8: APPLICATIONS

In this Applications Section, the format of the text has been altered. Answers for the problems appear beneath them rather than in the right-hand column. Your studying emphasis should be on learning the best procedures to follow with word problems. For that reason, once the procedure is learned a calculator may be used to complete the answer.

1

Carla's Bookstore has a mark-up of 40% on all of its books. This means that if a book costs Carla $15, she figures its selling price by finding 140% of 15. Find the selling price of a book that costs Carla $25.

Answer: 140% of 25 is found by multiplying 1.40 times 25 which is 35. Carla's selling price is $35 of which $25 (100% of 25) is the cost and $10 (40% of 25) is the mark-up.

2
A Buick agency sells it cars with a 7% mark-up. Find the selling price for a model that costs the agency $21,000.

Answer: The selling price is 107% of $21,000 which is found by multiplying 1.07 times 21,000. The answer is 22,470 which means the selling price is $22,470 of which $21,000 (100% of 21,000) is the cost and $1,470 (7% of 21,000) is the mark-up.

3
Pete's Appliances is having a sale on its stoves and offering a 15% discount. This means that a stove that normally sells for $250 will be priced at 85% (100% – 15%) of 250. Since .85 times 250 is 212.50, the sales price is $212.50 and the discount is $37.50 (15% of $250). Find the sale price of a $450 refrigerator which has a 20% discount.

Answer: The sales price is 80% (100% – 20%) of 450 or $360. The discount is $90 (20% of $450).

4
Judy's Dress Shop is having a sale on skirts which entails a 30% discount off of the normal price. Find the sales price of a skirt originally priced at $30.

Answer: The sales price is 70% (100% – 30%) of $30. Multiplying .70 times 30 gives 21 and the sales price is $21. The discount is $9 (30% of $30).

5
David is a realtor and makes a 3% commission on his sales. David sells a house for the price of $120,000. How much is David's commission on this sale?

Answer: The commission is 3% of $120,000 and is calculated as .03 times 120,000. The commission is $3,600.

6

A professional golfer announced that he was going to donate 25% of his winnings that week to the Heart Fund. He won the event! To fulfill his promise, how much of the first place money of $218,000 did he contribute to the Heart Fund?

Answer: 25% of $218,000 is found by multiplying .25 by 218,000. The calculation means that the Heart Fund will receive $54,500 and the golfer will keep $163,500 which is 75% of $218,000.

FEEDBACK UNIT 8 FOR APPLICATIONS

1. A vegetable stand owner figures he must have a 60% mark-up to pay for his costs and profit. If each head of lettuce costs him 70 cents, what should be his selling price for a head of lettuce?

2. A furniture store is having a sale on patio sets and offers a 30% discount. If the regular price for a patio set is $375, what would be its sales price?

3. An auctioneer believes he can sell the furnishings of an estate for $60,000. If he offers his services at a 12% commission, how much does he expect to make?

4. A beauty shop operator sells hair shampoo with an 80% mark-up. If a bottle of hair shampoo costs the operator $3.50, what should be its selling price?

5. Sam's Tires has a big sale and offers a 45% discount on the tires that are regularly $60 each. What is the sales price for one of these tires?

SUMMARY FOR THE ARITHMETIC OF THE RATIONAL NUMBERS

The following mathematical terms are crucial to an understanding of this chapter.

- Rational numbers
- Ratio
- Mixed number
- Improper fraction
- Numerator
- Denominator
- Integers as rational numbers
- Equal rational numbers
- Factors
- Simplest name for a rational number
- Simplifying rational numbers
- Evaluating a multiplication
- Cancelling
- Changing names of rational numbers
- Highest common factor, HCF
- Least common multiple, LCM
- Adding with a common denominator
- Opposites
- Reciprocals
- Division of rationals
- Complex fractions

The set of rational numbers was introduced in this chapter so that equations such as $7x = 3$ would have a number in the truth set.

The numerator of a rational number may be any integer, and the denominator of a rational number can be any integer except zero.

Every integer is in the set of rational numbers.

A method for writing rational numbers in their simplest form was shown.

A method was also shown to determine whether or not two rational numbers were equivalent.

The highest common factor and least common multiple were used to determine the lowest common denominator for evaluating addition expressions.

The highest common factor (HCF) is the largest factor that is common for two or more integers.

The least common multiple (LCM) is the smallest integer that can be divided evenly by two or more integers.

The evaluation of rational number expressions involving multiplication and addition was studied.

Subtraction of rational numbers was explained in terms of addition. Division of rational numbers was explained in terms of multiplication.

CHAPTER 7 MASTERY TEST

The following questions test the objectives of Chapter 7. Answers are at the back of the book. The number in parentheses which follows each problem indicates the unit in which it can be learned.

1. Which of the following is not a rational number?
   $\frac{3}{7}$ $\frac{-2}{5}$ $\frac{-6}{0}$ $\frac{-22}{-31}$ $\frac{0}{-9}$ (1)

2. Simplify. $\frac{-8}{-26}$ (3)

3. Simplify. $\frac{51}{-68}$ (3)

4. Simplify. $\frac{-15}{70}$ (3)

5. Find the highest common factor of 15 and 18. (5)

6. Find the least common multiple of 21 and 14. (5)

7. Evaluate. $\frac{2}{9} \bullet \frac{-4}{3}$ (4)

8. Evaluate. $\frac{-5}{7} \bullet 6$ (4)

9. Evaluate. $\frac{-7}{8} \bullet \frac{16}{21}$ (4)

10. Evaluate. $\frac{-8}{15} \bullet \frac{-15}{4}$ (4)

11. Evaluate. $\frac{5}{18} \bullet \frac{-18}{5}$ (4)

12. The opposite of 157 is ______. (6)

13. The reciprocal of -13 is ______. (7)

14. Evaluate. $\frac{-3}{8} + \frac{2}{3}$ (6)

15. Evaluate. $\frac{-7}{12} - \frac{-5}{18}$ (6)

16. Evaluate. $\frac{5}{18} + \frac{3}{24}$ (6)

17. Evaluate. $\frac{6}{35} - \frac{8}{49}$ (6)

18. Evaluate. $\frac{-3}{\frac{7}{9}}$ (7)

19. Evaluate. $\frac{\frac{-16}{3}}{-3}$ (7)

20. Evaluate. $\frac{\frac{-9}{14}}{\frac{-8}{21}}$ (7)

21. Sara's Shirt Outlet tries to mark-up its items 35%. If Sara pays $14 for a shirt, how much does she charge for it? (8)

22. Paul's Sport Shop is having a sale on soccer shoes and will give a 60% discount. If a pair of shoes was priced at $48, what will be the sales price? (8)

Chapter 8 Objectives

The following problems illustrate the objectives of this chapter. At this time you are not expected to know how to do these problems. However, if all these problems are thoroughly understood, proceed directly to the Chapter 8 Mastery Test. The number in parentheses which follows each problem indicates the unit in which it can be learned.

Simplify each of the following.

1. $4(3x - 7) - (x - 5)$ (3)
2. $-4(2x - 5) - 2(1 - 4x)$ (3)
3. $6(x - 9) + (5x - 1)$ (3)
4. $-(7 - 3x) + 6x - 16$ (3)
5. $(2x - 15) - (4x - 7)$ (3)
6. $\frac{2}{9}(3x - 10) - \left(\frac{5}{3}x - 3\right)$ (3)
7. $x^4 \bullet x^6$ (4)
8. $x^4 \bullet x$ (4)
9. $(x^5)^3$ (4)
10. $\frac{x^4}{x^9}$ (5)
11. $\frac{x^7}{x^3}$ (5)
12. $\frac{x^2}{x}$ (5)
13. $-(4x^2y)^3$ (4)
14. $(5xy^4)^3$ (4)
15. $-5x^3y^4 \bullet 4xy^2$ (4)
16. $\frac{-8x^2y^5}{4x^3y^5}$ (5)
17. $\frac{-x^5y^4}{x^5y^4}$ (5)
18. $\frac{-3xy^3}{-15xy^3}$ (5)

CHAPTER 8
THE ALGEBRA OF THE RATIONAL NUMBERS

UNIT 1: SIMPLIFYING ADDITION EXPRESSIONS

The following mathematical terms are crucial to an understanding of this unit.

Commutative Law of Addition
Associative Law of Addition
Simplifying Addition Expressions

1
The **Commutative Law of Addition** allows the order of addends to be reversed without altering the sum.

$\frac{1}{2}x + \frac{5}{9}$ is equivalent to $\frac{5}{9} + \frac{1}{2}x$

$\frac{3}{7}x + \frac{2}{3}$ is equivalent to ________

$\frac{2}{3} + \frac{3}{7}x$

2

The **Associative Law of Addition** allows the grouping of addends to be changed without altering the sum.

$\left(w+\frac{1}{2}\right)+\frac{-3}{5}$ is equivalent to $w+\left(\frac{1}{2}+\frac{-3}{5}\right)$

$\left[x+\frac{5}{3}\right]+(-2)$ is equivalent to ________

$x+\left[\frac{5}{3}+(-2)\right]$

3

To simplify $\left(t+\frac{-1}{2}\right)+\frac{1}{3}$, regroup the addends as:

$$t+\left(\frac{-1}{2}+\frac{1}{3}\right)$$

Evaluate. $\frac{-1}{2}+\frac{1}{3}$

$\frac{-1}{6}$

4

To **simplify** $\left[x+\frac{5}{3}\right]+(-2)$, the following steps are used:

$$\left[x+\frac{5}{3}\right]+(-2)$$
$$x+\left[\frac{5}{3}+(-2)\right]$$
$$x+\frac{-1}{3} \text{ or } x-\frac{1}{3}$$

Simplify. $\left(y+\frac{7}{8}\right)+3$

$y+\frac{31}{8}$

5

Complete the following simplification:

$\left(z+\frac{-3}{8}\right)+\frac{-5}{3}=z+\left(\frac{-3}{8}+\frac{-5}{3}\right)=$ ____________

$z+\frac{-49}{24}$ or $z-\frac{49}{24}$

6

Simplify. $\left(c+\frac{4}{3}\right)+\frac{-4}{3}$

c

7

Simplify. $\left(r+\frac{-6}{5}\right)+\frac{6}{5}$

r

8

Simplify. $\left(x+\frac{-5}{9}\right)+\frac{5}{9}$

x

9

To simplify $\left(\frac{2}{3}+x\right)+\frac{-3}{4}$, the Commutative Law of Addition is used to reorder the addends of $\left(\frac{2}{3}+x\right)$. This substitution gives $\left(x+\frac{2}{3}\right)+\frac{-3}{4}$. Complete the simplification:

$\left(\frac{2}{3}+x\right)+\frac{-3}{4}=\left(x+\frac{2}{3}\right)+\frac{-3}{4}=x+\left(\frac{2}{3}+\frac{-3}{4}\right)=$ ______

$x+\frac{-1}{12}$ or $x-\frac{1}{12}$

10
Complete the simplification:
$\left(\frac{4}{7}+x\right)+\frac{-2}{5}=\left(x+\frac{4}{7}\right)+\frac{-2}{5}=x+\left(\frac{4}{7}+\frac{-2}{5}\right)=$ _____

$x+\frac{6}{35}$

11
Simplify. $\left(\frac{9}{16}+z\right)+\frac{-9}{16}$

z

12
Simplify. $\left(c+\frac{-7}{8}\right)+\frac{1}{5}$

$c+\frac{-27}{40}$ or $c-\frac{27}{40}$

13
Simplify. $\left(r+\frac{3}{4}\right)+2$

$r+\frac{11}{4}$

14
Simplify. $\left(x+\frac{1}{3}\right)+\frac{1}{5}$

$x+\frac{8}{15}$

15
To simplify $\left[\frac{3}{4}x+\frac{5}{6}\right]+\frac{-2}{3}$, the following steps are used:

$$\left[\frac{3}{4}x+\frac{5}{6}\right]+\frac{-2}{3}$$
$$\frac{3}{4}x+\left[\frac{5}{6}+\frac{-2}{3}\right]$$
$$\frac{3}{4}x+\frac{1}{6}$$

Simplify. $\left(\frac{7}{8}x+\frac{2}{5}\right)-\frac{1}{4}$

$\frac{7}{8}x+\frac{3}{20}$

16
Complete the simplification:
$\left(\frac{2}{3}x-\frac{1}{3}\right)-\frac{2}{7}=\frac{2}{3}x+\left(\frac{-1}{3}-\frac{2}{7}\right)=$ _________

$\frac{2}{3}x-\frac{13}{21}$

17
Simplify. $\left(\frac{3}{5}x-\frac{5}{9}\right)+\frac{1}{3}$

$\frac{3}{5}x-\frac{2}{9}$

18
Simplify. $\left(\frac{3}{8}x+\frac{2}{3}\right)+\frac{3}{4}$

$\frac{3}{8}x+\frac{17}{12}$

19
Simplify. $\left(\frac{4}{7}+\frac{2}{5}x\right)-\frac{4}{7}$

$\frac{2}{5}x$

20
Simplify. $\left(\frac{5}{6}+\frac{7}{8}x\right)-\frac{1}{3}$

$\frac{7}{8}x+\frac{1}{2}$

21
Simplify. $\left(\frac{5}{8}+\frac{5}{2}x\right)-\frac{3}{4}$

$\frac{5}{2}x-\frac{1}{8}$

FEEDBACK UNIT 1

This quiz reviews the preceding unit. Answers are at the back of the book.

1. Simplify. $x + \left(\frac{3}{8} - \frac{1}{3}\right)$
2. Simplify. $\left(\frac{1}{2} + x\right) - \frac{1}{3}$
3. Simplify. $\frac{-5}{8} + \left(x + \frac{5}{8}\right)$
4. Simplify. $\left(\frac{2}{3} + x\right) + \frac{1}{6}$
5. Simplify. $x + \left(\frac{3}{4} - \frac{1}{3}\right)$
6. Simplify. $\left(r - \frac{2}{3}\right) + \frac{2}{3}$
7. Simplify. $\left(\frac{4}{5}x + \frac{2}{3}\right) - \frac{4}{7}$
8. Simplify. $\left(\frac{5}{6} + \frac{2}{3}x\right) - \frac{2}{5}$

UNIT 2: SIMPLIFYING MULTIPLICATION EXPRESSIONS

The following mathematical terms are crucial to an understanding of this unit.

Commutative Law of Multiplication
Associative Law of Multiplication
Multiplication Law of One
Multiplication Law of Negative One
Simplifying Multiplication Expressions

1

The **Commutative Law of Multiplication** allows the order of factors to be reversed without altering the product.

$\frac{1}{2}x \bullet \frac{5}{9}$ is equivalent to $\frac{5}{9} \bullet \frac{1}{2}x$

$\frac{3}{7}x \bullet \frac{2}{3}$ is equivalent to ________

$\frac{2}{3} \bullet \frac{3}{7}x$

2

The **Associative Law of Multiplication** allows the grouping of factors to be changed without altering the product.

$\frac{1}{2}\left(\frac{-3}{5}w\right)$ is equivalent to $\left(\frac{1}{2} \bullet \frac{-3}{5}\right)w$

$-4\left[\frac{5}{3}x\right]$ is equivalent to ________

$\left[-4 \bullet \frac{5}{3}\right]x$

3

To simplify $\left(\frac{5}{7}t\right) \cdot \frac{1}{3}$, reorder and regroup the factors.

$$\left(\frac{5}{7}t\right) \cdot \frac{1}{3} = \frac{1}{3} \cdot \left(\frac{5}{7}t\right) = \left(\frac{1}{3} \cdot \frac{5}{7}\right)t = \_\_\_\_\_$$

Complete the simplification. Multiply $\frac{1}{3}$ and $\frac{5}{7}$.

$\frac{5}{21}t$

4

To simplify $5 \cdot \left(\frac{3}{4}x\right)$, the Associative Law of Multiplication is used to write $\left(5 \cdot \frac{3}{4}\right)x$.

Simplify $\left(5 \cdot \frac{3}{4}\right)x$ by multiplying $\left(5 \cdot \frac{3}{4}\right)$.

$\frac{15}{4}x$

5

To simplify $\frac{-3}{8} \cdot \left(\frac{5}{9}x\right)$, the following steps are used:

$$\frac{-3}{8} \cdot \left(\frac{5}{9}x\right) = \left(\frac{-3}{8} \cdot \frac{5}{9}\right)x = \frac{-5}{24}x$$

Notice that cancellation was used in multiplying $\frac{-3}{8} \cdot \frac{5}{9}$.

Simplify. $-3 \cdot \left(\frac{5}{6}x\right)$

$\frac{-5}{2}x$

6

Complete the simplification:

$$\frac{-5}{12} \cdot \left(\frac{2}{3}x\right) = \left(\frac{-5}{12} \cdot \frac{2}{3}\right)x = \_\_\_\_\_$$

$\frac{-5}{18}x$

7

Simplify. $\frac{5}{6} \cdot (6x)$

5x

8

Simplify. $\frac{7}{2}\left(\frac{2}{7}z\right)$

1z or z

9

Simplify. $8\left(\frac{3}{8}x\right)$

3x

10

To simplify $\left(\frac{7}{4}x\right) \cdot 8$, the Commutative Law of Multiplication is used to write $8 \cdot \left(\frac{7}{4}x\right)$.

Complete the simplification:

$$\left(\frac{7}{4}x\right) \cdot 8 = 8 \cdot \left(\frac{7}{4}x\right) = \left(8 \cdot \frac{7}{4}\right)x = \_\_\_\_\_$$

14x

11

Simplify. $\left(\frac{-5}{4}x\right) \cdot \frac{4}{5}$

-1x or -x

12

Simplify. $\left(\frac{9}{14}x\right) \cdot 14$

9x

13
Simplify. $6 \cdot \left(\frac{-5}{4}x\right)$

$\frac{-15}{2}x$

14
Simplify $\frac{3}{4} \cdot \left(\frac{3}{2}x\right)$ which is equivalent to $\frac{3}{4}\left(\frac{3}{2}x\right)$.

$\frac{9}{8}x$

15
Simplify. $\frac{5}{6}\left(\frac{4}{5}x\right)$

$\frac{2}{3}x$

16
Simplify. $\frac{3}{4}\left(\frac{-8}{9}x\right)$

$\frac{-2}{3}x$

17
Simplify. $\left(\frac{7}{5}x\right) \cdot \frac{-10}{3}$

$\frac{-14}{3}x$

18
The open sentence 1x = x becomes true when x is replaced by any rational number. Each of the following statements is true.

$1 \cdot \frac{9}{4} = \frac{9}{4}$ $\quad 1 \cdot \frac{-5}{3} = \frac{-5}{3}$ $\quad 1 \cdot \frac{-7}{33} = \frac{-7}{33}$ $\quad 1 \cdot \frac{6}{7} = \frac{6}{7}$

Is there any rational number replacement for x that will make 1x = x false?

No

19
1x = x becomes a true statement for all rational-number replacements of x. This is the **Multiplication Law of One**. 1z = z becomes a ______ (true or false) statement for all rational-number replacements of z.

true

20
The simplification of $\frac{5}{6}\left(\frac{6}{5}x\right)$ is 1x or _____.

x

21
Simplify. $\frac{-5}{9}\left(\frac{-9}{5}x\right)$

x

22
-x means the opposite of the replacement for x. The open sentence -1x = -x becomes true when x is replaced by any rational number. Each of the following statements is true.

$-1 \cdot \frac{9}{4} = \frac{-9}{4}$ $\quad -1 \cdot \frac{-5}{3} = \frac{5}{3}$ $\quad -1 \cdot \frac{-7}{33} = \frac{7}{33}$ $\quad -1 \cdot \frac{6}{7} = \frac{-6}{7}$

Is there any rational number replacement for x that will make -1x = -x false?

No

23
Whenever -1 is multiplied by a rational number, the product is the opposite of that rational number. $-1y = -y$ becomes a ______ (true or false) statement for all rational-number replacements for y.

true

24
The open sentence $-1x = -x$ is the **Multiplication Law of Negative One**. Is -1 multiplied by a rational number always going to produce the opposite of the rational number?

Yes

25
Simplify. $-1(\frac{9}{8}x)$

$\frac{-9}{8}x$

26
Simplify. $-1(\frac{-3}{56}x)$

$\frac{3}{56}x$

27
Simplify. $\frac{-5}{4}(\frac{4}{5}x)$

$-x$

28
Simplify. $\frac{10}{9}(\frac{-9}{10}x)$

$-x$

29
Simplify. $(\frac{2}{3}x) \cdot \frac{3}{2}$

x

30
Simplify. $(\frac{4}{7}x) \cdot \frac{-7}{4}$

$-x$

FEEDBACK UNIT 2

This quiz reviews the preceding unit. Answers are at the back of the book.

Simplify each of the following.

1. $\frac{2}{3} \cdot (\frac{3}{4}x)$
2. $(\frac{8}{5}x) \cdot \frac{-4}{3}$
3. $\frac{5}{6} \cdot (\frac{-7}{10}x)$
4. $\frac{-7}{9} \cdot (\frac{-2}{3}x)$
5. $\frac{4}{5} \cdot (\frac{7}{8}x)$
6. $(\frac{-8}{9}x) \cdot \frac{3}{4}$
7. $\frac{3}{4}(\frac{4}{3}a)$
8. $\frac{-6}{7}(\frac{-7}{6}w)$
9. $8(\frac{3}{8}x)$
10. $(\frac{5}{6}x) \cdot 6$

UNIT 3: SIMPLIFYING OPEN EXPRESSIONS

The following mathematical terms are crucial to an understanding of this unit.

Distributive Law of Multiplication over Addition
Like terms
Coefficients

1
To evaluate $10 \bullet \left(\frac{3}{5} + \frac{-2}{3}\right)$, the parentheses indicate that the addition is done first. To evaluate $10 \bullet \frac{3}{5} + 10 \bullet \frac{-2}{3}$ the lack of parentheses indicates that the ___________ is done first.

multiplication

2
The evaluations of $10 \bullet \left(\frac{3}{5} + \frac{-2}{3}\right)$ and $10 \bullet \frac{3}{5} + 10 \bullet \frac{-2}{3}$ are shown below.

$10 \bullet \left(\frac{3}{5} + \frac{-2}{3}\right)$	$10 \bullet \frac{3}{5} + 10 \bullet \frac{-2}{3}$
$10 \bullet \left(\frac{-1}{15}\right)$	$6 + \frac{-20}{3}$
$\frac{-10}{15}$	$\frac{18}{3} + \frac{-20}{3}$
$\frac{-2}{3}$	$\frac{-2}{3}$

The evaluations are the same. Is this a coincidence?

No

3
The evaluations of $4\left(\frac{5}{7} + \frac{2}{3}\right)$ and $4\left(\frac{5}{7}\right) + 4\left(\frac{2}{3}\right)$ will be the same because the two expressions are of the forms a(b + c) and ab + ac. The **Distributive Law of Multiplication over Addition** guarantees that a(b + c) is equivalent to _________.

ab + ac

4
Will $\frac{3}{4}\left(\frac{5}{6} - \frac{1}{8}\right)$ and $\frac{3}{4}\left(\frac{5}{6}\right) + \frac{3}{4}\left(\frac{-1}{8}\right)$ have the same evaluation?

Yes

5
Will $\frac{2}{3}\left(\frac{9}{5} + \frac{1}{4}\right)$ and $\frac{2}{3}\left(\frac{9}{5}\right) + \frac{2}{3}\left(\frac{1}{4}\right)$ have the same evaluation?

Yes

6
The open sentence $x(y + z) = xy + xz$ becomes a true statement for all rational number replacements of x, y, and z. The Distributive Law of Multiplication over Addition guarantees that $5(x + 6)$ is equivalent to _________.

$5 \bullet x + 5 \bullet 6$ or $5x + 30$

7
By the Distributive Law of Multiplication over Addition,

$$4\left(x + \frac{2}{5}\right) \text{ is equivalent to } 4x + 4 \bullet \frac{2}{5} \text{ or } 4x + \frac{8}{5}$$

Similarly, $-5\left(z + \frac{4}{5}\right)$ is equivalent to __________

$-5z + (-5) \bullet \frac{4}{5}$ or $-5z - 4$

8
The expression $2x + 5x$ can be simplified by using the Distributive Law of Multiplication over Addition, as shown in the following steps:

$$2x + 5x$$
$$(2 + 5)x$$
$$7x$$

Use the Distributive Law of Multiplication over Addition to simplify $7y + 5y$.

$(7 + 5)\,y = 12y$

9
In the addition expression $18x - 5x$, the addends $18x$ and $-5x$ are **like terms**. Simplify $18x - 5x$ by combining its like terms.

$(18 - 5)x = 13x$

10
The addition expression $13z + 4z$ contains the like terms $13z$ and $4z$. Simplify $13z + 4z$.

$(13 + 4)z = 17z$

11
In the addition expression $4x - 10x$, the numbers 4 and -10 are **coefficients**. Simplify $4x - 10x$ by adding the coefficients of its like terms.

$-6x$

12
What are the coefficients in the addition expression $-5x + 7x$?

-5 and 7

13
Simplify $5x + 3x$ by adding like terms.

$8x$

14
Simplify $9x + 4x$ by adding like terms.

$13x$

15
Simplify. $2x + 4x$

$6x$

16
Simplify. $9x - 7x$

$9x + (-7x) = 2x$

17
Simplify. $3x - 11x$

$-8x$

18
Simplify. $-2x - 5x$

$-7x$

19
To simplify $\frac{3}{4}x - \frac{2}{3}x$, the following steps are used:

$$\frac{3}{4}x - \frac{2}{3}x$$
$$\frac{3}{4}x + \frac{-2}{3}x$$
$$\left(\frac{3}{4} + \frac{-2}{3}\right)x$$
$$\frac{1}{12}x$$

Simplify. $\frac{7}{8}x - \frac{1}{5}x$

$\frac{27}{40}x$

20
Simplify. $2x - \frac{5}{6}x$

$\left(\frac{2}{1} + \frac{-5}{6}\right)x = \frac{7}{6}x$

21
Simplify. $\frac{2}{3}x - \frac{1}{4}x$

$\frac{5}{12}x$

22
Simplify. $\frac{7}{6}y - \frac{1}{2}y$

$\frac{4}{6}y = \frac{2}{3}y$

23
To remove the parentheses from the expression $5\left(x + \frac{2}{3}\right)$, separately multiply both x and $\frac{2}{3}$ by 5.

$$5\left(x + \frac{2}{3}\right)$$
$$5 \bullet x + 5 \bullet \frac{2}{3}$$
$$5x + \frac{10}{3}$$

Remove the parentheses from $2\left(y + \frac{3}{5}\right)$.

$2y + \frac{6}{5}$

24
Remove the parentheses from $7\left(x + \frac{4}{3}\right)$.

$7x + \frac{28}{3}$

25
Remove the parentheses from $5\left(x + \frac{2}{7}\right)$.

$5x + \frac{10}{7}$

26
Remove the parentheses from $4\left(x - \frac{2}{3}\right)$.
[Note: $\left(x - \frac{2}{3}\right) = \left(x + \frac{-2}{3}\right)$]

$4x - \frac{8}{3}$

27
Remove the parentheses from $5\left(x - \frac{3}{2}\right)$.

$5x - \frac{15}{2}$

28
Remove the parentheses from $-3\left(x - \frac{5}{4}\right)$.

$-3x - \frac{-15}{4} = -3x + \frac{15}{4}$

29
To remove the parentheses from -5(2x + 3), multiply both 2x and 3 by -5.

$$-5(2x + 3)$$
$$-5 \cdot 2x + (-5 \cdot 3)$$
$$-10x + (-15)$$
$$-10x - 15$$

Remove the parentheses from the expression -6(3x + 1).

$-18x + (-6) = -18x - 6$

30
Complete the simplification:
$8(3x - 2) = 8 \cdot 3x + 8 \cdot -2 =$ ______

$24x - 16$

31
Complete the simplification:
$-5(2x - 4) = -5 \cdot 2x + (-5) \cdot -4 =$ ______

$-10x + 20$

32
Remove the parentheses from the expression (7x + 5).
[Note: (7x + 5) is equivalent to 1 • (7x + 5)]

$7x + 5$

33
Remove the parentheses. $\left(\frac{3}{5}x - \frac{7}{8}\right)$

$\frac{3}{5}x - \frac{7}{8}$

34
Remove the parentheses. -(2x + 3)
[Note: -(2x + 3) is equivalent to -1(2x + 3)]

$-2x - 3$

35
Remove the parentheses. -(3x – 5)

$-3x + 5$

36
Remove the parentheses. -5(x – 2)

$-5x + 10$

37
Remove the parentheses. $\frac{1}{4}(8x - 3)$

$2x - \frac{3}{4}$

38
Remove the parentheses. $-2(7x + 5)$ — $-14x - 10$

39
Remove the parentheses. $\left(8 - \frac{17}{5}x\right)$ — $8 - \frac{17}{5}x$

40
Remove the parentheses. $7(5 - 2x)$ — $35 - 14x$

41
Remove the parentheses. $\frac{2}{5}(10 - 5x)$ — $4 - 2x$

42
Remove the parentheses. $-(3x - 4)$ — $-3x + 4$

43
Remove the parentheses. $-3(4x + 1)$ — $-12x - 3$

44
Remove the parentheses. $-6(5 - x)$ — $-30 + 6x$

45
Remove the parentheses. $-4(5 - x)$ — $-20 + 4x$

46
Remove the parentheses. $2(5 - 3x)$ — $10 - 6x$

47
Remove the parentheses. $-(7 - x)$ — $-7 + x$

48
Remove the parentheses. $-8(4 - 2x)$ — $-32 + 16x$

49
Remove the parentheses. $(6 - 9x)$ — $6 - 9x$

50
To simplify $5 + 2x + 6x + 11$, group the like terms.

$$5 + 2x + 6x + 11$$
$$(2x + 6x) + (5 + 11)$$
$$8x + 16$$

Simplify $7 + 4x + 3x + 5$ by first grouping like terms. — $7x + 12$

51
Simplify $3x + 9 - 2x - 2$ by grouping $3x$ with $-2x$ and 9 with -2. — $x + 7$

52
Simplify $8x - 13 + 2 - 6x$ by grouping the addends properly.

$2x - 11$

53
Simplify. $5x - 3 + 8 + 8x$

$13x + 5$

54
Simplify. $2x + 3 - 5x - 3$

$-3x$

55
Simplify. $2 - 4x + 7x - 6$

$3x - 4$

56
Simplify. $9 + 6x - 5 - 11x$

$-5x + 4$

57
To simplify $8 - 3(x + 5)$, first remove the parentheses by multiplying $(x + 5)$ by -3.

$$8 - 3(x + 5)$$
$$8 - 3x - 15$$
$$-3x - 7$$

Simplify $5 - 2(x + 1)$ by first multiplying $(x + 1)$ by -2.

$-2x + 3$

58
Simplify $6 - 4(1 - 3x)$ by first multiplying $(1 - 3x)$ by -4.

$2 + 12x$

59
Simplify $2x - (4 - 3x)$ by first multiplying $(4 - 3x)$ by -1.

$5x - 4$

60
Simplify $5x + 2(x - 7)$ by first multiplying $(x - 7)$ by 2.

$7x - 14$

61
Simplify. $4 + (7x - 3)$

$7x + 1$

62
Simplify. $3x - 2(x + 5)$

$x - 10$

63
Simplify. $4 + 7(x - 2)$

$7x - 10$

64
Simplify. $-3 - 5(1 - 2x)$ — $10x - 8$

65
Simplify. $9x - 3(2x + 1)$ — $3x - 3$

66
Simplify. $2x - (5x - 1)$ — $-3x + 1$

67
To simplify $2(x - 1) + 3(4 - 2x)$, first remove both pairs of parentheses.

$$2(x - 1) + 3(4 - 2x)$$
$$2x - 2 + 12 - 6x$$
$$(2x - 6x) + (-2 + 12)$$
$$-4x + 10$$

Complete the simplification:

$5(2x - 1) + (4x - 3) = 10x - 5 + 4x - 3 =$ ______ — $14x - 8$

68
Complete the simplification:

$7(3 - x) - (4x - 2) = 21 - 7x - 4x + 2 =$ ______ — $-11x + 23$

69
Simplify. $(5x - 7) - (3x + 5)$ — $2x - 12$

70
Simplify. $4(2x - 3) + 3(x + 4)$ — $11x$

71
Simplify. $2x - 7(x + 3) + 5$ — $-5x - 16$

72
Simplify. $3(2x - 5) + 2(5x - 2)$ — $16x - 19$

73
To simplify $\frac{2}{5}x + 5 + 2x$, the following steps are used:

$$\frac{2}{5}x + 5 + 2x$$
$$\left(\frac{2}{5}x + 2x\right) + 5$$
$$\left(\frac{2}{5} + 2\right)x + 5$$
$$\frac{12}{5}x + 5$$

Simplify $\frac{3}{4}x + 7 + 5x$ by first grouping like terms. — $\frac{23}{4}x + 7$

74
Complete the simplification:
$\frac{1}{4}x + 5 + \frac{2}{3}x = \left(\frac{1}{4}x + \frac{2}{3}x\right) + 5 = \left(\frac{1}{4} + \frac{2}{3}\right)x + 5 =$ ______

$\frac{11}{12}x + 5$

75
Complete the simplification:
$\frac{2}{7}x + \frac{1}{3} + \frac{4}{5}x - \frac{1}{4} = \left(\frac{2}{7}x + \frac{4}{5}x\right) + \left(\frac{1}{3} - \frac{1}{4}\right) =$ ______

$\frac{38}{35}x + \frac{1}{12}$

76
Simplify. $\frac{5}{6}x + 4 + \frac{5}{6}x$

$\frac{5}{3}x + 4$

77
Simplify. $5 + \frac{9}{10}x + \frac{3}{4}$

$\frac{9}{10}x + \frac{23}{4}$

78
Complete the simplification below where the parentheses are removed as the first step.
$2x + 5\left(\frac{2}{5}x + \frac{1}{3}\right) = 2x + 2x + \frac{5}{3} =$ ______________

$4x + \frac{5}{3}$

79
Complete the simplification below where the parentheses are removed as the first step.
$3\left(2x - \frac{1}{7}\right) + \left(3x + \frac{5}{6}\right) = 6x - \frac{3}{7} + 3x + \frac{5}{6} =$ ______

$9x + \frac{17}{42}$

80
Complete the simplification below where the parentheses are removed as the first step.
$-(2x + 5) + 3(x - 4) = -2x - 5 + 3x - 12 =$ ______

$x - 17$

81
Simplify. $3\left(2x - \frac{1}{4}\right) - 2(x + 3)$

$4x - \frac{27}{4}$

82
Simplify. $-5(2x + 3) + 3(x - 4)$

$-7x - 27$

83
Simplify $3\left(\frac{1}{4}x + \frac{1}{5}\right) - 2\left(\frac{1}{3}x + \frac{1}{5}\right)$ by first removing parentheses.

$\frac{1}{12}x + \frac{1}{5}$

84
Simplify. $\frac{1}{3}\left(\frac{1}{4} - \frac{2}{5}x\right) + \frac{5}{8}x$

$\frac{59}{120}x + \frac{1}{12}$

85
Simplify. $\frac{1}{2}\left(\frac{5}{6}x - \frac{1}{3}\right) + \frac{3}{4}x$

$\frac{7}{6}x - \frac{1}{6}$

FEEDBACK UNIT 3

This quiz reviews the preceding unit. Answers are at the back of the book.

1. Are the evaluations the same for $15(\frac{3}{5} + \frac{1}{3})$ and $15 \cdot \frac{3}{5} + 15 \cdot \frac{1}{3}$?
2. Remove the parentheses. $7(\frac{3}{7}x + 4)$
3. Remove the parentheses. $\frac{2}{3}(\frac{1}{4}x - 2)$
4. Simplify. $3x + 8x$
5. Simplify. $-3a + 8a$
6. Simplify. $5(4x + 7)$
7. Simplify. $-3(7a + 4)$
8. Simplify. $3x + 2(x - 5)$
9. Simplify. $4 - (3x - 7)$
10. Simplify. $7(3x - 2) - 5(2x - 7)$
11. Simplify. $\frac{3}{4}x + 5 + 4x - \frac{2}{5}$
12. Simplify. $\frac{1}{2}(6x - 8) + 2x$

UNIT 4: MULTIPLICATION USING EXPONENTS

The following mathematical terms are crucial to an understanding of this unit.

Exponents	Factor
Base	Power expression
Raising a power to a power	Coefficient

1

Exponents often appear in multiplication expressions, but the **exponents** are not factors. Is $5^2 = 2 \cdot 5$ true?

No; $5^2 = 5 \cdot 5$

2

In the open expression x^2, the exponent 2 shows that x is a **factor** two times in multiplication. For the open expression x^5, the exponent shows that x is to be a factor ______ times in the multiplication.

5

3
The **base** of x^2 is x and x^2 is equivalent to xx.
x^3 is equivalent to ______. 	xxx

4
y^2 is a **power expression** and is equivalent to ______. 	yy

5
The power expression x^2 is equivalent to xx.
The power expression y^3 is equivalent to ______. 	yyy

6
The base of x^4 is x and x^4 is equivalent to ______. 	xxxx

7
yyyyy is equivalent to y^5.
xxx is equivalent to the power expression ______. 	x^3

8
zzzz is equivalent to the power expression ______. 	z^4

9
aaa is equivalent to ______. 	a^3

10
xxx is equivalent to x^3
xxxxx is equivalent to ______. 	x^5

11
$a^2 \bullet a^4$ is the multiplication of two power expressions and
$a^2 \bullet a^4$ is equivalent to aa • aaaa.
aaaaaa is equivalent to the power expression ______. 	a^6

12
The multiplication of power expressions with the same base can be simplified to a single power expression.

$c^4 \bullet c^3 = cccc \bullet ccc = ccccccc = c^7$

$x \bullet x^2 = x \bullet xx = xxx =$ ______ 	x^3

13
The multiplication of power expressions with the same base can be simplified.

$w^2 \bullet w^5 = ww \bullet wwwww = wwwwwww = w^7$

$z \bullet z^4 = z \bullet zzzz = zzzzz =$ ______ 	z^5

14
To simplify $x^5 \cdot x^8$ notice that the only factors are x's.
Is 5 a factor of $x^5 \cdot x^8$?

No, x is a factor 5 times in the expression x^5.

15
x^2 means that x is used as a factor two times. x^4 means that x is used as a factor four times. In the multiplication expression $x^2 \cdot x^4$, how many times is x used as a factor?

6

16
y^2 means that y is used as a factor two times.
y^3 means that y is used as a factor three times.
$y^2 \cdot y^3$ means that y is used as a factor ______ times.

5

17
In $x^2 \cdot x^7$, how many factors of x are used?

9

18
In $x \cdot x^9$, how many factors of x are used?

10

19
$c^6 \cdot c^2 = cccccc \cdot cc = c^8$
$c^4 \cdot c = cccc \cdot c =$ ______

c^5

20
$d^2 \cdot d^3 = dd \cdot ddd =$ ______

d^5

21
$m^5 \cdot m^2 = mmmmm \cdot mm =$ ______

m^7

22
The expression $x^4 \cdot x^5$ can be simplified to x^9 because $x^4 \cdot x^5 = xxxx \cdot xxxxx = x^9$. Simplify $x^2 \cdot x^8$

x^{10}

23
To simplify $x^5 \cdot x^3$, the number of times x is used as a factor is needed. Recall that addition is counting. Consequently, $x^5 \cdot x^3$ can be simplified by adding the exponents.

$$x^5 \cdot x^3 = x^{5+3} = x^8$$

Simplify. $x^5 \cdot x^{11}$

$x^{5+11} = x^{16}$

24
Simplify. $y^4 \cdot y^7$

$y^{4+7} = y^{11}$

25
Simplify $a^5 \cdot a$ by recognizing that $a = a^1$.

$a^{5+1} = a^6$

26
Simplify. $a^3 \cdot a$

a^4

27
Simplify. $x^9 \cdot x^5$

x^{14}

28
When power expressions with the same base are multiplied, their exponents are added.

$k^a \cdot k^b = k^{a+b}$ $\quad y^a \cdot y^b = y^{a+b}$ $\quad r^a \cdot r^b = r^{a+b}$

Can the multiplication of $x^3 \cdot y^5$ be simplified by adding the exponents?

No, the bases are different.

29
To simplify $x^{11} \cdot x^{24}$, the exponents are not multiplied. The exponents show how many times x is to be used as a factor. What is the total number of x's used as factors in $x^{11} \cdot x^{24}$?

35

30
There are 35 factors of x in the open expression $x^{11} \cdot x^{24}$. Write a power expression to show 35 factors of x.

x^{35}

31
$x^{11} \cdot x^{24}$ is simplified to x^{35} because the exponents show there are 35 factors of x. Simplify $x^{18} \cdot x^3$

x^{21}

32
$x^2 \cdot x^5 \cdot x^3$ is a multiplication of three power expressions. Is the base the same for all three power expressions?

Yes, it is x.

33
In the open expression $x^2 \cdot x^5 \cdot x^3$ what is the total number of factors of x?

10

34
x^{10} is an open expression showing that x is a factor 10 times. Can $x^2 \cdot x^5 \cdot x^3$ be simplified to x^{10}?

Yes

35
How many factors of x are in the open expression $x^5 \cdot x^9 \cdot x^2$?

16

36
Simplify. $x^5 \cdot x^9 \cdot x^2$

x^{16}

37
Simplify. $x \cdot x^8 \cdot x^3$

x^{12}

38
Simplify. $x^2 \cdot x^2 \cdot x^3$

x^7

39
Simplify. $x^4 \cdot x^3 \cdot x$

x^8

40
Simplify. $x^6 \cdot x \cdot x^7$

x^{14}

41
$3 \cdot 7^2$ means $3 \cdot 7 \cdot 7$
$2 \cdot (-5)^3$ means $2 \cdot -5 \cdot -5 \cdot -5$
$5x^3$ means 5xxx. $3x^2$ means ______

3xx

42
In $5x^2$, the 5 is a **coefficient** of x^2.
$5x^2$ means ______

5xx

43
$-3x^4$ has -3 as a coefficient and means -3xxxx.
$-4y^3$ means ______

-4yyy

44
To simplify $2x^5 \cdot 5x^3$, separately group the coefficients and the power expressions.

$$2x^5 \cdot 5x^3 = (2 \cdot 5)(x^5 \cdot x^3) = 10x^8$$

Simplify. $3x^2 \cdot 7x^6$

$(3 \cdot 7)(x^2 \cdot x^6) = 21x^8$

45
Complete the following simplification:

$$6x^2 \cdot 5x^3 = (6 \cdot 5)(x^2 \cdot x^3) = \_\_\_\_\_\_$$

$30x^5$

46
Simplify. $2x^5 \cdot 8x$

$16x^6$

47
Simplify. $-3x^2 \cdot 2x^{11}$

$-6x^{13}$

48
Simplify $5x^9 \bullet -x^8$ by recognizing that $-x^8 = -1x^8$ — $-5x^{17}$

49
Simplify. $-9x \bullet -4x$ — $36x^2$

50
Simplify. $-x^5 \bullet 4x^3$ — $-4x^8$

51
To simplify $4x^2y \bullet 3x^2y^7$, form separate groups of the coefficients and each base.

$$4x^2y \bullet 3x^2y^7 = (4 \bullet 3)(x^2x^2)(yy^7) = 12x^4y^8$$

Simplify. $5x^7y^3 \bullet 2x^5y^8$ — $10x^{12}y^{11}$

52
Complete the following simplification:

$$-2x^4y \bullet 3x^2y^3 = (-2 \bullet 3)(x^4x^2)(yy^3) = \_\_\_\_\_\_$$

$-6x^6y^4$

53
Complete the following simplification:

$$5xy^4 \bullet -x^7y^3 = (5 \bullet -1)(xx^7)(y^4y^3) = \_\_\_\_\_\_$$

$-5x^8y^7$

54
Simplify. $5x^5y^3 \bullet -4xy^7$ — $-20x^6y^{10}$

55
Simplify. $-x^2y^2 \bullet -3xy^3$ — $3x^3y^5$

56
Simplify. $7x^8y^6 \bullet 3x^2y^5$ — $21x^{10}y^{11}$

57
Simplify. $8x^4y^3 \bullet -3xy^6$ — $-24x^5y^9$

58
Simplify. $-5x^8y^2 \bullet -4x^2y^5$ — $20x^{10}y^7$

59
Simplify. $7xy \bullet 6x^3y^5$ — $42x^4y^6$

60
Simplify. $9x^4y \bullet -x^3y^2$ — $-9x^7y^3$

61
The first step in simplifying the multiplication expression $5x^2y^5z^3 \cdot -2xy^4z^3$ is to regroup the factors as follows:

$$5x^2y^5z^3 \cdot -2xy^4z^3$$
$$(5 \cdot -2)(x^2x)(y^5y^4)(z^3z^3)$$
$$-10x^3y^9z^6$$

Can the result $-10x^3y^9z^6$ be simplified by adding the exponents?

No, the bases are different.

62
To simplify $2x^5yz^2 \cdot -3x^2yz^9$, notice that there are three different bases for the power expressions. Each must be grouped separately.

$$2x^5yz^2 \cdot -3x^2yz^9$$
$$(2 \cdot -3)(x^5x^2)(yy)(z^2z^9)$$
$$-6x^7y^2z^{11}$$

Simplify. $4x^5y^2z^3 \cdot 6x^2y^3z^6$

$24x^7y^5z^9$

63
Complete the following simplification:

$5xy^3z \cdot -x^3y^5z^4 = (5 \cdot -1)(xx^3)(y^3y^5)(zz^4) =$ ______

$-5x^4y^8z^5$

64
Complete the following simplification:

$3x^5yz^4 \cdot -8x^2y^5z^3 = (3 \cdot -8)(x^5x^2)(yy^5)(z^4z^3) =$ ______

$-24x^7y^6z^7$

65
Simplify. $-3xy^2z \cdot -5x^2y^3z^3$

$15x^3y^5z^4$

66
Simplify. $6xy^4z^7 \cdot -x^3y^3z^3$

$-6x^4y^7z^{10}$

67
Simplify. $4x^2y^2z^3 \cdot -8xy^2z^3$

$-32x^3y^4z^6$

68
Simplify. $-5x^4y^2z^3 \cdot -10xyz^3$

$50x^5y^3z^6$

69
Simplify. $x^4y^2z \cdot 18xyz^5$

$18x^5y^3z^6$

70
Simplify. $4xy^4z^2 \cdot 4xy^4z^2$

$16x^2y^8z^4$

71
$2x^3$ means 2xxx because the base of the exponent is x.
$(2x)^3$ means $2x \cdot 2x \cdot 2x$ because the base of the exponent is _____.

2x

72
$(-3x^5y)^2$ means $-3x^5y \cdot -3x^5y$. The base of the exponent 5 is x, but the base of the exponent 2 is ______.

$-3x^5y$

73
$(3x^2)^4$ means $3x^2 \cdot 3x^2 \cdot 3x^2 \cdot 3x^2$
$(2x^6)^3$ means ______

$2x^6 \cdot 2x^6 \cdot 2x^6$

74
To simplify $(-2x^5y)^3$, use the base of the exponent 3 three times.

$$(-2x^5y)^3$$
$$(-2x^5y)(-2x^5y)(-2x^5y)$$
$$(-2 \cdot -2 \cdot -2)(x^5x^5x^5)(yyy)$$
$$-8x^{15}y^3$$

Simplify. $(3x^4y^3)^2$

$9x^8y^6$

75
Complete the following simplification:

$$(9xy^2z^5)^2$$
$$(9xy^2z^5)(9xy^2z^5)$$
$$(9 \cdot 9)(xx)(y^2y^2)(z^5z^5)$$

____

$81x^2y^4z^{10}$

76
Complete the following simplification:

$$(-x^3y^4z^2)^3$$
$$(-x^3y^4z^2)(-x^3y^4z^2)(-x^3y^4z^2)$$
$$(-1 \cdot -1 \cdot -1)(x^3x^3x^3)(y^4y^4y^4)(z^2z^2z^2)$$

____

$-x^9y^{12}z^6$

77
Complete the following simplification:

$$(xy^5z^2)^3$$
$$(xy^5z^2)(xy^5z^2)(xy^5z^2)$$
$$(xxx)(y^5y^5y^5)(z^2z^2z^2)$$

____

$x^3y^{15}z^6$

78
Simplify. $(5x^2y^3z)^2$

$25x^4y^6z^2$

79
Simplify. $(2x^5y^2z^3)^3$

$8x^{15}y^6z^9$

80
Simplify. $(-4x^4y^7z)^2$

$16x^8y^{14}z^2$

81
A shorter method for simplifying $(-4x^2yz^4)^3$ uses each factor of the base $-4x^2yz^4$ three times.

$$(-4x^2yz^4)^3$$
$$(-4)^3(x^2)^3(y)^3(z^4)^3$$
$$(-64)(x^6)(y^3)(z^{12})$$
$$-64x^6y^3z^{12}$$

Notice that $(x^2)^3$ is equivalent to x^6. The exponents were multiplied! This is because $(x^2)^3 = x^2x^2x^2 = x^6$. Show why $(z^4)^3 = z^{12}$.

$(z^4)^3 = z^4z^4z^4 = z^{12}$

82
In **raising a power to a power**, as in $(x^3y^2z^5)^4$, the exponents can be multiplied.

$$(x^3y^2z^5)^4$$
$$(x^3)^4(y^2)^4(z^5)^4$$
$$(x^{3 \cdot 4})(y^{2 \cdot 4})(z^{5 \cdot 4})$$
$$x^{12}y^8z^{20}$$

Simplify. $(xy^2z^5)^3$

$x^{1 \cdot 3}y^{2 \cdot 3}z^{5 \cdot 3} = x^3y^6z^{15}$

83
Simplify. $(-7x^8y^3z^4)^2$

$(-7)^2x^{16}y^6z^8 = 49x^{16}y^6z^8$

84
In multiplying power expressions with the same base, exponents are added. $x^4 \cdot x^4 = x^{4+4} = x^8$ In raising a power to a power, exponents are multiplied.

$(x^2)^3 = x^{2 \cdot 3} = x^6$

$x^a \cdot x^b = x^{a+b}$ and $(x^a)^b =$ _________

x^{ab}

85
If there is any doubt about the correct method for simplifying an exponent expression, rely upon the meaning of an exponent.

$x^4 \cdot x^5$ means xxxx • xxxxx which means x^9

$(x^4)^2$ means $(x^4)(x^4)$ which means xxxx • xxxx or x^8

Simplify. $(-3x^5)^2$

$9x^{10}$

86
Simplify. $(2x^3)^5$

$32x^{15}$

87
Simplify. $-2x^6y^4 \bullet 5x^7y^3$

$-10x^{13}y^7$

88
Simplify. $(-2x^2y)^3$

$-8x^6y^3$

89
Simplify. $(-4a^3b^2c)^4$

$256a^{12}b^8c^4$

90
Simplify. $-(4a^3b^2c)^4$

$-1(256a^{12}b^8c^4)$ which equals $-256a^{12}b^8c^4$

FEEDBACK UNIT 4

This quiz reviews the preceding unit. Answers are at the back of the book.

Simplify each of the following.

1. $x^7 \bullet x^4$
2. $x^4 \bullet x^5$
3. $x^7 \bullet x^9$
4. $3x^4 \bullet -2x^2$
5. $-x^5 \bullet 5x^2$
6. $5x^2y \bullet -3x^3y$
7. $-6x^3y \bullet 4x^6y^4$
8. $(2x^3y)^3$
9. $(-3x^2y)^4$
10. $(2xy^3)^2$
11. $-(5x^2y)^4$
12. $(-5x^2y)^4$
13. $-(2x)^3$
14. $(-2x)^3$

UNIT 5: DIVISION USING EXPONENTS

The following mathematical term is crucial to an understanding of this unit.

Dividing power expressions

1

In the fraction $\frac{x^2}{x^3}$ there are two factors of x in the numerator and ______ factors of x in the denominator.

3

2

To simplify $\frac{x^2}{x^3}$, two factors of x are divided out of both the numerator and denominator.

$$\frac{x^2}{x^3} = \frac{x \cdot x}{x \cdot x \cdot x} = \frac{\cancel{x} \cdot \cancel{x}}{\cancel{x} \cdot \cancel{x} \cdot x} = \frac{1}{x}$$

Simplify $\frac{x^3}{x^4}$ by dividing three factors of x out of both the numerator and denominator.

$$\frac{x \cdot x \cdot x}{x \cdot x \cdot x \cdot x} = \frac{1}{x}$$

3

To simplify $\frac{x^5}{x^2}$, two factors of x are divided out of both the numerator and denominator.

$$\frac{x^5}{x^2} = \frac{xxxxx}{xx} = \frac{\cancel{xx} \cdot xxx}{\cancel{xx}} = \frac{x^3}{1} = x^3$$

Simplify $\frac{x^5}{x^3}$ by dividing out three factors of x.

$$\frac{xxx \cdot xx}{xxx} = \frac{x^2}{1} = x^2$$

4

Because $\frac{x^5}{x^5} = 1$ and $x^9 = x^5 \cdot x^4$, the following simplification can be written with exponents.

$$\frac{x^9}{x^5} = \frac{x^5 \cdot x^4}{x^5} = \frac{\cancel{x^5} \cdot x^4}{\cancel{x^5}} = \frac{x^4}{1} = x^4$$

Simplify $\frac{x^{12}}{x^4}$ by dividing out all possible factors of x.

$$\frac{x^4 \cdot x^8}{x^4} = \frac{x^8}{1} = x^8$$

5

Complete the simplification: $\frac{x^4}{x} = \frac{x \cdot x^3}{x} =$ ______

$\frac{x^3}{1} = x^3$

6

Complete the simplification: $\frac{x^2}{x^6} = \frac{x^2}{x^2 \cdot x^4} =$ ______

$\frac{1}{x^4}$

7

Complete the simplification: $\frac{x^5}{x^8} = \frac{x^5}{x^5 \cdot x^3} =$ ______

$\frac{1}{x^3}$

8

The simplification of $\frac{x^9}{x^2}$ can be shortened using the fact that each power expression will have

2 less (subtraction) factors of x.

$$\frac{x^9}{x^2} = \frac{x^{9-2}}{1} = \frac{x^7}{1} = x^7$$

Simplify. $\frac{x^{10}}{x^7}$

$\frac{x^{10-7}}{1} = \frac{x^3}{1} = x^3$

9

Simplify. $\frac{x^8}{x^4}$

$\frac{x^{8-4}}{1} = \frac{x^4}{1} = x^4$

10

Simplify. $\frac{x^7}{x^6}$

x

11

Simplify. $\frac{x^{10}}{x^5}$

x^5

12

To simplify $\frac{x^2}{x^8}$, notice that the denominator has more factors of x.

$$\frac{x^2}{x^8} = \frac{1}{x^{8-2}} = \frac{1}{x^6}$$

Simplify. $\frac{x^{12}}{x^{14}}$

$\frac{1}{x^{14-12}} = \frac{1}{x^2}$

13

Simplify. $\frac{x^5}{x^8}$

$\frac{1}{x^{8-5}} = \frac{1}{x^3}$

14

Simplify. $\frac{x^8}{x^{10}}$

$\frac{1}{x^2}$

15

Simplify. $\frac{x^3}{x^6}$

$\frac{1}{x^3}$

16

Simplify. $\frac{x^4}{x^5}$

$\frac{1}{x}$

17

In **dividing power expressions** with the same base the exponents can be subtracted.

$$\frac{x^5}{x^2} = \frac{x^{5-2}}{1} = x^3 \quad \text{and} \quad \frac{x^3}{x^9} = \frac{1}{x^{9-3}} = \frac{1}{x^6}$$

When k is greater than m, $\frac{x^k}{x^m} = \frac{x^{k-m}}{1} = x^{k-m}$

When k is less than m, $\frac{x^k}{x^m} =$ _______

$\frac{1}{x^{m-k}}$

18

Simplify. $\frac{x^4}{x^7}$

$\frac{1}{x^3}$

19

To simplify $\frac{-8x^5}{12x^3}$, separately simplify the coefficients and the power expressions.

$$\frac{-8x^5}{12x^3} = \frac{-8}{12} \bullet \frac{x^5}{x^3} = \frac{-2}{3} \bullet \frac{x^2}{1} = \frac{-2x^2}{3}$$

Complete the following simplification:

$$\frac{9x^7}{12x^4} = \frac{9}{12} \bullet \frac{x^7}{x^4} = \underline{\qquad}$$

$\frac{3x^3}{4}$

20

Complete the following simplification:

$$\frac{8x^4}{4x} = \frac{8}{4} \bullet \frac{x^4}{x} = \underline{\qquad}$$

$2x^3$

21

Simplify. $\frac{3x^5}{9x^2}$

$\frac{1}{3} \bullet \frac{x^3}{1} = \frac{x^3}{3}$ or $\frac{1}{3}x^3$

22

Simplify. $\frac{-4x^8}{6x^4}$

$\frac{-2}{3} \bullet \frac{x^4}{1} = \frac{-2x^4}{3}$ or $\frac{-2}{3}x^4$

23
Simplify. $\frac{7x^4}{21x}$

$\frac{x^3}{3}$ or $\frac{1}{3}x^3$

24
Simplify. $\frac{-10x^9}{-5x^2}$

$2x^7$

25
Simplify. $\frac{24x}{3x^2}$

$\frac{8}{x}$

26

To simplify $\frac{-15x^2y^5z^3}{10x^6y^5z}$, the coefficients and different bases are separated. The simplification is done as follows:

$$\frac{-15}{10} \bullet \frac{x^2}{x^6} \bullet \frac{y^5}{y^5} \bullet \frac{z^3}{z}$$

$$\frac{-3}{2} \bullet \frac{1}{x^4} \bullet \frac{1}{1} \bullet \frac{z^2}{1} = \frac{-3z^2}{2x^4}$$

Complete the following simplification.

$$\frac{12x^5y^7z^3}{8x^6y^5z^3} = \frac{12}{8} \bullet \frac{x^5}{x^6} \bullet \frac{y^7}{y^5} \bullet \frac{z^3}{z^3} = \text{\_\_\_\_\_\_\_\_\_\_\_\_}$$

$\frac{3}{2} \bullet \frac{1}{x} \bullet \frac{y^2}{1} \bullet \frac{1}{1} = \frac{3y^2}{2x}$

27

To simplify $\frac{10xy^7z^3}{15x^2y^3z}$, the following steps are used:

$$\frac{10}{15} \bullet \frac{x}{x^2} \bullet \frac{y^7}{y^3} \bullet \frac{z^3}{z}$$

$$\frac{2}{3} \bullet \frac{1}{x} \bullet \frac{y^4}{1} \bullet \frac{z^2}{1} = \frac{2y^4z^2}{3x}$$

Complete the following simplification.

$$\frac{3x^2y^5z^7}{6x^4y^2z^2} = \frac{3}{6} \bullet \frac{x^2}{x^4} \bullet \frac{y^5}{y^2} \bullet \frac{z^7}{z^2} = \text{\_\_\_\_\_\_\_\_\_\_\_\_}$$

$\frac{1}{2} \bullet \frac{1}{x^2} \bullet \frac{y^3}{1} \bullet \frac{z^5}{1} = \frac{y^3z^5}{2x^2}$

28
Complete the following simplification.

$$\frac{6x^5y^2z^8}{3x^2y^6z^4} = \frac{6}{3} \bullet \frac{x^5}{x^2} \bullet \frac{y^2}{y^6} \bullet \frac{z^8}{z^4} = \text{\_\_\_\_\_\_\_\_\_\_\_\_}$$

$\frac{2}{1} \bullet \frac{x^3}{1} \bullet \frac{1}{y^4} \bullet \frac{z^4}{1} = \frac{2x^3z^4}{y^4}$

29
Simplify. $\dfrac{3x^3y^7z^2}{6xy^3z^4}$

$\dfrac{1}{2} \bullet \dfrac{x^2}{1} \bullet \dfrac{y^4}{1} \bullet \dfrac{1}{z^2} = \dfrac{x^2y^4}{2z^2}$

30
Simplify. $\dfrac{7x^8y}{x^2y^5z^2}$

$\dfrac{7}{1} \bullet \dfrac{x^6}{1} \bullet \dfrac{1}{y^4} \bullet \dfrac{1}{z^2} = \dfrac{7x^6}{y^4z^2}$

31
Simplify. $\dfrac{12x^5y^2z^4}{2x^3y^2z^6}$

$\dfrac{6}{1} \bullet \dfrac{x^2}{1} \bullet \dfrac{1}{1} \bullet \dfrac{1}{z^2} = \dfrac{6x^2}{z^2}$

32
Simplify. $\dfrac{-5x^3y^5z^6}{15x^3y^2z^6}$

$\dfrac{-1}{3} \bullet \dfrac{1}{1} \bullet \dfrac{y^3}{1} \bullet \dfrac{1}{1} = \dfrac{-y^3}{3}$

33
Simplify. $\dfrac{40x^3y^2z^5}{8x^2y^5z^2}$

$\dfrac{5xz^3}{y^3}$

34
Simplify. $\dfrac{-20xy^3z^5}{4x^3y^3z^3}$

$\dfrac{-5z^2}{x^2}$

35
Simplify. $\dfrac{-8x^2y^3z^6}{12x^2y^5z^4}$

$\dfrac{-2z^2}{3y^2}$

36
Simplify. $\dfrac{-32x^3yz^5}{-8x^3yz^2}$

$4z^3$

37
Simplify. $\dfrac{8x^3yz^2}{8x^3yz^2}$

1

38
Simplify. $\dfrac{14x^2y^3z}{9xy^5z^3}$

$\dfrac{14x}{9y^2z^2}$

39
Simplify. $\dfrac{-10x^2y^3z}{10x^2y^3z}$

-1

40
Simplify. $\dfrac{x^3y^2}{-3xy^5}$

$\dfrac{-x^2}{3y^3}$

41
Simplify. $\frac{-5a^3c}{15a^3}$

$\frac{-c}{3}$

42
Simplify. $\frac{5a^2b^3c}{-5a^2b^2c}$

-b

FEEDBACK UNIT 5

This quiz reviews the preceding unit. Answers are at the back of the book.

Simplify each of the following.

1. $\frac{x^5}{x^2}$
2. $\frac{x^7}{x^4}$
3. $\frac{x^2}{x^6}$
4. $\frac{x^5}{x^5}$
5. $\frac{x^3}{x^8}$
6. $\frac{x^{10}}{x^6}$
7. $\frac{x^4}{x^8}$
8. $\frac{2x^3}{8x}$
9. $\frac{-5x^2y}{5xy}$
10. $\frac{-15x^8y}{10x^5y^7}$
11. $\frac{8x^5y^3z}{16x^2y^3z^5}$

UNIT 6: APPLICATIONS

In this Applications Section, the format of the text has been altered. Answers for the problems appear beneath them rather than in the right-hand column. Your studying emphasis should be on learning the best procedures to follow with word problems. For that reason, once the procedure is learned a calculator may be used to complete the answer.

1

An inch, a foot, and a meter are examples of linear measurements. With linear measurements, distances or lengths are stated. Find the perimeter of the figure shown at the right.

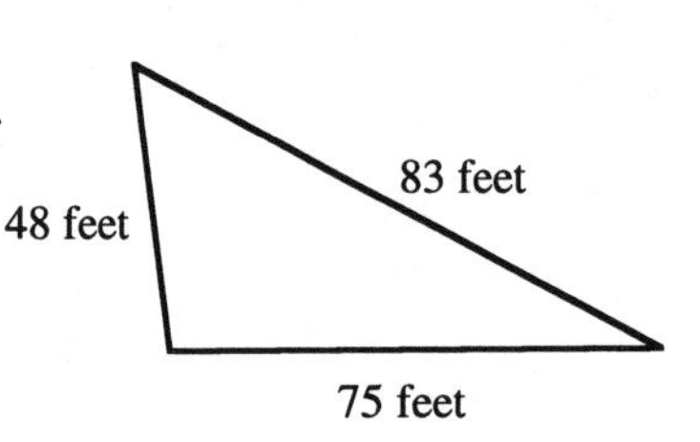

Answer: The figure is a triangle and its perimeter is the sum of the lengths of its sides. Those lengths are linear measurements. The perimeter is found by evaluating 75 + 83 + 48. The perimeter is 206 feet.

2

Find the perimeter of the figure shown at the right.

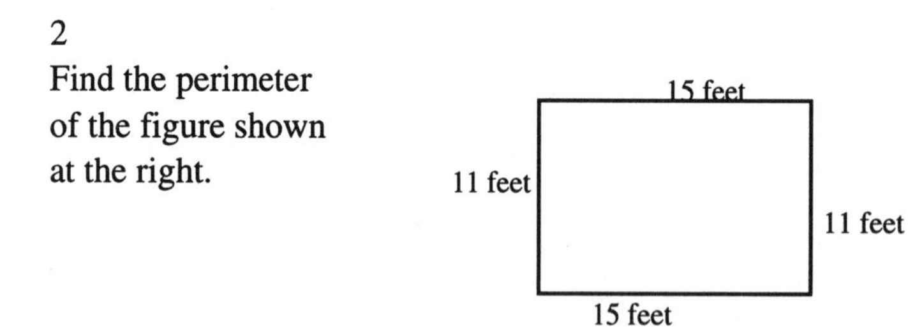

Answer: The figure is a rectangle and its perimeter can be found as the sum of the lengths of its sides. Another method for finding the perimeter of a rectangle is through the use of the formula P = 2L + 2W where L is the perimeter's length and W is its width.

$$P = 11 + 15 + 11 + 15 \quad \text{or} \quad P = 2 \cdot 15 + 2 \cdot 11$$

Using either of the methods shown above, the perimeter is 52 feet.

3

A square inch, square foot, or square meter are units used to measure the area or space of a plane figure (one that can be drawn on a piece of paper). The figure at the right shows a rectangle with length 8 inches and width 5 inches. The small gray square represents one square inch. Find the area, in square inches, of the rectangle using the formula A = LW.

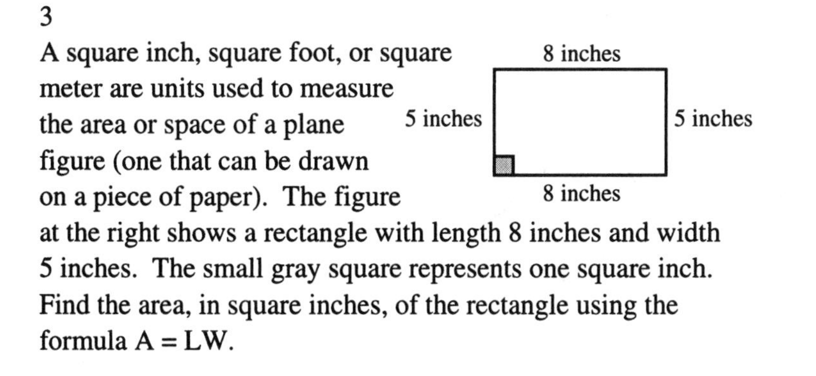

Answer: The area is 40 square inches. Notice that when two linear measurements are multiplied, the product will be in square units.

4

Find the area of the triangle shown at the right using the formula $A = \frac{1}{2}bh$ where b stands for the base of the triangle and h stands for the height to that base.

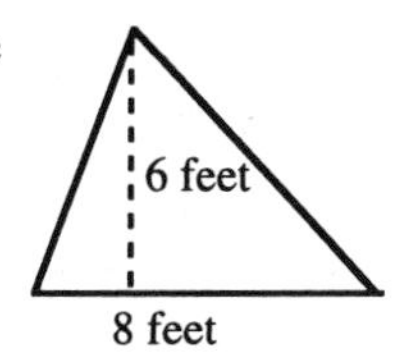

Answer: The area is 24 square feet. In the formula $A = \frac{1}{2}bh$, b and h represent linear measures; when they are multiplied the result is in square units.

5

The amount of air in a room is measured in cubic units. The figure at the right shows a box that is 10 inches in length, 6 inches wide, and 4 inches high. The small gray box is one cubic inch. Find the volume, in cubic units, of the box using the formula V = LWH.

4 inches
6 inches
10 inches

Answer: The volume is 240 cubic inches. Notice that when three linear measures are multiplied, the product is in cubic units.

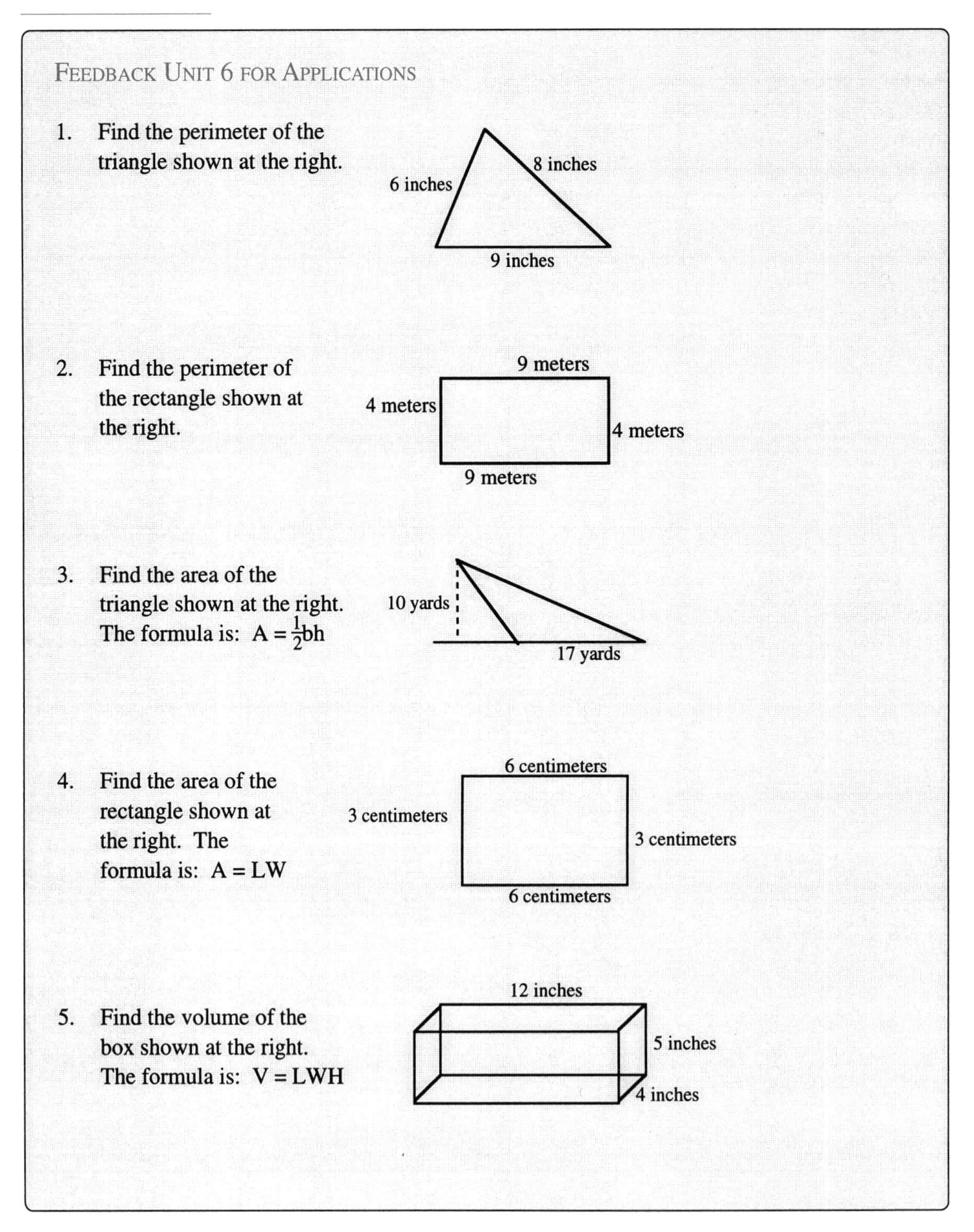

FEEDBACK UNIT 6 FOR APPLICATIONS

1. Find the perimeter of the triangle shown at the right.

2. Find the perimeter of the rectangle shown at the right.

3. Find the area of the triangle shown at the right. The formula is: $A = \frac{1}{2}bh$

4. Find the area of the rectangle shown at the right. The formula is: $A = LW$

5. Find the volume of the box shown at the right. The formula is: $V = LWH$

SUMMARY FOR THE ALGEBRA OF THE RATIONAL NUMBERS

The following mathematical terms are crucial to an understanding of this unit.

Commutative Law of Addition
Simplifying Addition Expressions
Associative Law of Multiplication
Multiplication Law of Negative One
Distributive Law of Multiplication over Addition
Factors
Power expression
Raising a power to a power
Associative Law of Addition
Commutative Law of Multiplication
Multiplication Law of One
Simplifying Multiplication Expressions
Like terms
Exponents
Base
Coefficient
Dividing power expressions

The set of rational numbers has the following laws, which make it an extremely important set of numbers.

a. The Commutative and Associative Laws of Addition.
b. The Commutative and Associative Laws of Multiplication.
c. The Addition Law of Zero.
d. Every rational number has an opposite.
e. The Multiplication Law of One.
f. Every rational number except zero has a reciprocal.
g. The Distributive Law of Multiplication over Addition.
h. The Multiplication Law of Negative One.
i. The sum of any rational number and its opposite is zero.
j. The product of a rational number and its reciprocal is one.

The properties of the set of rational numbers make it possible to find a unique solution for every equation such as $2x + 7 = 4$ and $5x + 8 = 24 - 3x$.

Three Laws of Exponents were explained:

1. In multiplying like bases the exponents are added.

$$x^3 \cdot x^7 = x^{3+7} = x^{10}$$

2. When the base is an exponent expression, the exponents are multiplied.

$$(x^3)^7 = x^{3 \cdot 7} = x^{21}$$

3. In division of like bases the exponents are subtracted.

$$\frac{x^8}{x^3} = x^{8-3} = x^5 \text{ or } \frac{x^5}{x^7} = \frac{1}{x^{7-5}} = \frac{1}{x^2}$$

CHAPTER 8 MASTERY TEST

The following questions test the objectives of Chapter 8. Answers are at the back of the book. The number in parentheses which follows each problem indicates the unit in which it can be learned.

Simplify each of the following.

1. $3(2x - 5) - (x - 4)$ (3)
2. $-5(3x - 7) - 2(x - 1)$ (3)
3. $3(x - 3) - (x + 5)$ (3)
4. $-(4 - 3x) + 8 - 3x$ (3)
5. $\frac{2}{3}(6x - 5) - \frac{1}{2}(x - 3)$ (3)
6. $\frac{4}{7}(3x - 14) + \frac{5}{8}(4x - 1)$ (3)
7. $x^3 \bullet x^5$ (4)
8. $x \bullet x^7$ (4)
9. $(x^3)^6$ (4)
10. $\frac{x^2}{x^5}$ (5)
11. $\frac{x^7}{x^3}$ (5)
12. $\frac{x}{x^2}$ (5)
13. $-8x^3 \bullet 7x^9$ (4)
14. $-3x \bullet x^5$ (4)
15. $-(2x^2y)^4$ (4)
16. $(3x^3y^2)^3$ (4)
17. $-3x^4y^3 \bullet 8xy^7$ (4)
18. $\frac{-5x^5y}{-10x^2y^2}$ (5)
19. $\frac{-x^6y^7}{x^6y^7}$ (5)
20. $\frac{x^3y^2}{x^2y^3}$ (5)
21. Find the perimeter of the rectangle shown below. (6)

22. Find the area of the rectangle shown below. (6)

CHAPTER 9 OBJECTIVES

The following problems illustrate the objectives of this chapter. At this time you are not expected to know how to do these problems. However, if all these problems are thoroughly understood, proceed directly to the Chapter 9 Mastery Test. The number in parentheses which follows each problem indicates the unit in which it can be learned.

Using the set of rational numbers, find a solution for each of the following equations.

1. $x + \frac{3}{4} = \frac{5}{6}$ (1)

2. $x - \frac{7}{8} = \frac{1}{5}$ (1)

3. $\frac{5}{7}x = \frac{2}{3}$ (2)

4. $\frac{-3}{5}x = \frac{7}{10}$ (2)

5. $4x + 2 = 13$ (3)

6. $-5x + 7 = 18$ (3)

7. $3x - 2 = 5x + 17$ (3)

8. $x + 9 = 3x - 23$ (3)

9. $5 - 3x = 19 - x$ (3)

10. $-6 + 3x = 5x - 21$ (3)

11. $\frac{7}{3}x - 1 = \frac{-2}{7}$ (3)

12. $3(x - 4) - 5 = x + 1$ (3)

13. $7 - (5x - 6) = 6$ (3)

14. $2(4x - 1) + 8 = -(4x + 3) - 4$ (3)

15. $\frac{-2}{5}x - \frac{2}{3} = -2$ (3)

16. $\frac{4}{5}x - \frac{1}{4} = x + \frac{3}{10}$ (3)

17. $\frac{-3}{7}x - \frac{5}{21} = \frac{1}{3}$ (3)

Solve and check the following equations.

18. $\frac{2}{9}x - \frac{1}{3} = \frac{5}{6}$ (4)

19. $\frac{1}{3} - 2x = \frac{-3}{4}x + \frac{5}{12}$ (4)

20. $\frac{7}{8}x - \frac{2}{3} = \frac{5}{6}x - \frac{1}{2}$ (4)

CHAPTER 9

SOLVING EQUATIONS WITH THE RATIONAL NUMBERS

UNIT 1: FINDING SOLUTIONS BY ADDING OPPOSITES

The following mathematical terms are crucial to an understanding of this unit.

Equivalent equations Solve

1
The only solution for $x + 6 = 13$ is 7. When 7 replaces x, $7 + 6 = 13$ is true. Find the solution for $x + 9 = 15$ by asking:
What number can be added to 9 to give 15?

6, $6 + 9 = 15$ is true

2
Find the solution for $x - 7 = 2$ by asking:
What number can be added to -7 to give 2? 9

3
Find the solution for $x - 3 = 10$ by asking:
What number can be added to -3 to give 10? 13

4
Find the solution for $3x = 15$ by asking:
What number can be multiplied by 3 to give 15? 5, $3 \cdot 5 = 15$ is true

5
Find the solution for $-4x = 24$ by asking:
What number can be multiplied by -4 to give 24? -6

6
Find the solution for $-5x = -35$ by asking:
What number can be multiplied by -5 to give -35? 7

7
Two numbers, like 13 and -13, are opposites because their sum is zero. To find the solution of $5x + 13 - 13 = 45$, first simplify the left side of the equation.

$$5x + 13 - 13 = 45$$
$$5x + 0 = 45$$
$$5x = 45$$

What is the solution for $5x + 13 - 13 = 45$? 9

8
The opposite of -3 is 3 because their sum is zero. Find the solution of $7x - 3 + 3 = 42$ by first adding -3 and 3.

$$7x - 3 + 3 = 42$$
$$7x + 0 = 42$$
$$7x = 42$$

What is the solution for $7x - 3 + 3 = 42$? 6

9
The equations $7x - 3 + 3 = 42$ and $7x = 42$ have the same solution. Do $5x + 13 - 13 = 45$ and $5x = 45$ have the same solution? Yes, solution is 9

10
$2x + 7 - 7 = 14$ and $2x = 14$ are **equivalent equations** because they both have 7 as their solution. Are $4x = 20$ and $x = 5$ equivalent equations? Yes, solution is 5

11
Any rational number may be added to both sides of an equation to generate an equivalent equation. To find the solution for $4x - 3 = 17$, first add 3 to both sides of the equation.

$$\begin{array}{rcl} 4x - 3 &=& 17 \\ +3 && +3 \\ \hline 4x + 0 &=& 20 \\ 4x &=& 20 \end{array}$$

What is the solution for $4x - 3 = 17$?

5

12
To find the solution for $2x + 7 = 19$, first add -7 to both sides of the equation.

$$\begin{array}{rcl} 2x + 7 &=& 19 \\ -7 && -7 \\ \hline 2x + 0 &=& 12 \\ 2x &=& 12 \end{array}$$

What is the solution for $2x + 7 = 19$?

6

13
Add -9 to both sides of the equation $6x + 9 = 45$. What equivalent equation is obtained?

$6x = 36$

14
For the equation $3x + 6 = 15$, what number is added to both sides of the equation to obtain the equivalent equation $3x = 9$?

-6

15
To find the solution for $2x + 7 = 13$, add the opposite of 7 to each side of the equation. Find the solution for $2x + 7 = 13$.

$2x = 6$, solution is 3

16
Find the solution for $3x - 8 = -2$ by first adding the opposite of -8 to both sides of the equation.

$3x = 6$, solution is 2

17
Find the solution for $-7x + 3 = -11$ by first adding the opposite of 3 to both sides of the equation.

$-7x = -14$, solution is 2

18
Find the solution for $3x + 4 = 19$ by generating an equivalent equation.

$3x = 15$, solution is 5

19
Solve $-4x + 3 = 23$ by generating an equivalent equation.
[Note: "Solve" means to find any rational number that is a solution.]

$-4x = 20$, solution is -5

20
Solve $7x - 2 = 12$ by generating an equivalent equation.

$7x = 14$, solution is 2

21
Solve. $2x - 5 = 27$

16

22
Solve. $-6x + 13 = 1$

2

23
Solve $15 = 4x + 3$. If it makes it easier, write $4x + 3 = 15$.

3

24
Solve $20 = 6 - 7x$. If it makes it easier, write $20 = -7x + 6$.

-2

25
Solve. $5x - 3 = 27$

6

26
The opposite of $\frac{15}{11}$ is $\frac{-15}{11}$. The sum of any rational number and its opposite is ______.

zero

27
Every rational number has an opposite.
What is the opposite of $\frac{9}{4}$?

$\frac{-9}{4}$

28
What is the opposite of $\frac{-7}{15}$?

$\frac{7}{15}$

29
To find the solution for $x + \frac{5}{7} = \frac{3}{4}$, the opposite of $\frac{5}{7}$ is added to both sides of the equation.

$$\begin{aligned} x + \tfrac{5}{7} &= \tfrac{3}{4} \\ \left(x + \tfrac{5}{7}\right) - \tfrac{5}{7} &= \left(\tfrac{3}{4}\right) - \tfrac{5}{7} \\ x + \left(\tfrac{5}{7} - \tfrac{5}{7}\right) &= \left(\tfrac{3}{4} - \tfrac{5}{7}\right) \\ x + 0 &= \left(\tfrac{21}{28} - \tfrac{20}{28}\right) \\ x &= \tfrac{1}{28} \end{aligned}$$

Therefore, the solution of $x + \frac{5}{7} = \frac{3}{4}$ is ______.

$\frac{1}{28}$

30

To find the solution for $x - \frac{1}{3} = \frac{2}{9}$, first add the opposite of $\frac{-1}{3}$ to both sides of the equation.

$$\begin{aligned} x - \tfrac{1}{3} &= \tfrac{2}{9} \\ \left(x - \tfrac{1}{3}\right) + \tfrac{1}{3} &= \left(\tfrac{2}{9}\right) + \tfrac{1}{3} \\ x + \left(\tfrac{-1}{3} + \tfrac{1}{3}\right) &= \left(\tfrac{2}{9} + \tfrac{1}{3}\right) \\ x + 0 &= \left(\tfrac{2}{9} + \tfrac{3}{9}\right) \\ x &= \tfrac{5}{9} \end{aligned}$$

The solution for $x - \frac{1}{3} = \frac{2}{9}$ is ______.

$\frac{5}{9}$

31

Find the solution for $x + \frac{3}{4} = \frac{2}{3}$ by adding the opposite of $\frac{3}{4}$ to both sides of the equation.

$x + 0 = \frac{2}{3} - \frac{3}{4}$

$x = \frac{-1}{12}$

32

Find the solution for $x - \frac{5}{9} = \frac{3}{2}$ by adding the opposite of $\frac{-5}{9}$ to both sides of the equation.

$x + 0 = \frac{3}{2} + \frac{5}{9}$

$x = \frac{37}{18}$

33

Solve. $x - \frac{6}{7} = \frac{1}{3}$

$x + 0 = \frac{1}{3} + \frac{6}{7}$

$x = \frac{25}{21}$

34

Solve. $x - \frac{5}{8} = \frac{11}{8}$

2

35

Solve. $x + \frac{1}{2} = \frac{1}{5}$

$\frac{-3}{10}$

36

Solve. $x - \frac{3}{4} = \frac{1}{2}$

$\frac{5}{4}$

37

Solve. $x + 3 = \frac{2}{3}$

$\frac{-7}{3}$

38

Solve. $x - \frac{3}{4} = 3$

$\frac{15}{4}$

39
Solve. $x - \frac{1}{2} = \frac{5}{8}$

$\frac{9}{8}$

40
Solve. $x + \frac{3}{5} = \frac{2}{3}$

$\frac{1}{15}$

41
Solve. $x - \frac{2}{7} = 1$

$\frac{9}{7}$

FEEDBACK UNIT 1

This quiz reviews the preceding unit. Answers are at the back of the book.

Solve each of the following equations.

1. $5x = -20$
2. $-3x = 18$
3. $4x + 7 = 3$
4. $5 - 3x = -7$
5. $x - \frac{3}{4} = \frac{5}{6}$
6. $x + \frac{1}{8} = \frac{2}{7}$
7. $x - 2 = \frac{1}{4}$
8. $x + \frac{5}{9} = 4$

UNIT 2: FINDING SOLUTIONS BY MULTIPLYING BY RECIPROCALS

1
Two numbers are **reciprocals** when their product is one.

$\frac{6}{7} \cdot \frac{7}{6} = 1 \qquad \frac{11}{8} \cdot \frac{8}{11} = 1 \qquad \frac{-4}{3} \cdot \frac{-3}{4} = 1$

What is the reciprocal of 7?

$\frac{1}{7}$, because $7 \cdot \frac{1}{7} = 1$

2
If both sides of $7x = 21$ are multiplied by $\frac{1}{7}$, the new equation will be **equivalent**. Are $7x = 21$ and $\frac{1}{7} \cdot 7x = \frac{1}{7} \cdot 21$ equivalent equations?

Yes, 3 is the solution for both

3

If both sides of $3x = 6$ are multiplied by $\frac{1}{3}$, the new equation will be equivalent.

$$3x = 6$$
$$\frac{1}{3} \bullet 3x = \frac{1}{3} \bullet 6$$
$$1x = 2$$
$$x = 2$$

What is the solution for $3x = 6$?

2

4

To solve $7x = 17$, both sides of the equation are multiplied by the reciprocal of 7.

$$7x = 17$$
$$\frac{1}{7} \bullet 7x = \frac{1}{7} \bullet 17$$
$$1x = \frac{17}{7}$$
$$x = \frac{17}{7}$$

What is the solution for $7x = 17$?

$\frac{17}{7}$

5

To solve $-5x = 11$, an equivalent equation is generated by multiplying both sides by $\frac{-1}{5}$.

$$-5x = 11$$
$$\frac{-1}{5} \bullet -5x = \frac{-1}{5} \bullet 11$$
$$1x = \frac{-11}{5}$$
$$x = \frac{-11}{5}$$

What is the solution for $-5x = 11$?

$\frac{-11}{5}$

6

Find the solution for $3x = 23$ by multiplying both sides by the reciprocal of 3.

$\frac{1}{3} \bullet 3x = \frac{1}{3} \bullet 23$

Solution is $\frac{23}{3}$

7

Find the solution for $-4x = 5$ by multiplying both sides by the reciprocal of -4.

$\frac{-1}{4} \bullet -4x = \frac{-1}{4} \bullet 5$

Solution is $\frac{-5}{4}$

8
Find the solution for $8x = -7$ by generating an equivalent equation.

$1x = \frac{-7}{8}$
Solution is $\frac{-7}{8}$

9
Find the solution for $-9x = -4$ by generating an equivalent equation.

$1x = \frac{4}{9}$
Solution is $\frac{4}{9}$

10
To solve the equation $\frac{3}{4}x = 2$, the following steps are used.

$$\begin{aligned} \frac{3}{4}x &= 2 \\ \frac{4}{3} \bullet \frac{3}{4}x &= \frac{4}{3} \bullet 2 \\ 1x &= \frac{8}{3} \\ x &= \frac{8}{3} \end{aligned}$$

What is the solution for $\frac{3}{4}x = 2$?

$\frac{8}{3}$

11
If both sides of $\frac{7}{8}x = 3$ are multiplied by the reciprocal of $\frac{7}{8}$, what equivalent equation is obtained?

$\frac{8}{7} \bullet \frac{7}{8}x = \frac{8}{7} \bullet 3$ or $1x = \frac{24}{7}$

12
What is the solution for $\frac{7}{8}x = 3$ or, equivalently, $1x = \frac{24}{7}$?

$\frac{24}{7}$

13
If both sides of $\frac{-5}{9}x = 2$ are multiplied by the reciprocal of $\frac{-5}{9}$, what equivalent equation is obtained?

$\frac{-9}{5} \bullet \frac{-5}{9}x = \frac{-9}{5} \bullet 2$
or $1x = \frac{-18}{5}$

14
What is the solution for $\frac{-5}{9}x = 2$ or, equivalently, $1x = \frac{-18}{5}$?

$\frac{-18}{5}$

15

The **coefficient** of x in $\frac{4}{3}x = 8$ is $\frac{4}{3}$. Multiply both sides of the equation by the reciprocal of the coefficient of x and find the equivalent equation.

$\frac{3}{4} \cdot \frac{4}{3}x = \frac{3}{4} \cdot 8$
or $1x = 6$

16

Multiply both sides of $\frac{-5}{6}x = 7$ by the reciprocal of the coefficient of x and find the equivalent equation.

$\frac{-6}{5} \cdot \frac{-5}{6}x = \frac{-6}{5} \cdot 7$
or $1x = \frac{-42}{5}$

17

Multiply both sides of $\frac{2}{3}x = \frac{5}{7}$ by the reciprocal of the coefficient of x and find the equivalent equation.

$\frac{3}{2} \cdot \frac{2}{3}x = \frac{3}{2} \cdot \frac{5}{7}$
or $1x = \frac{15}{14}$

18

Find a solution for $\frac{7}{3}x = 2$ by multiplying both sides of the equation by the reciprocal of the coefficient of x.

$\frac{6}{7}$

19

Find a solution for $\frac{-5}{7}x = 4$ by multiplying both sides of the equation by the reciprocal of the coefficient of x.

$\frac{-28}{5}$

20

Find a solution for $\frac{-5}{3}x = -4$ by multiplying both sides of the equation by the reciprocal of the coefficient of x.

$\frac{12}{5}$

21

Find a solution for $\frac{2}{3}x = 7$.

$\frac{21}{2}$

22

Solve $\frac{-3}{5}x = 4$
[Note: "Solve" means to find the solution.]

$\frac{-20}{3}$

23
Solve. $\frac{9}{4}x = 3$

$\frac{4}{3}$

24
Solve. $\frac{7}{9}x = 7$

9

25
Solve. $\frac{-4}{5}x = 2$

$\frac{-5}{2}$

26
Solve. $\frac{2}{7}x = -5$

$\frac{-35}{2}$

27
Solve. $\frac{-2}{5}x = -3$

$\frac{15}{2}$

28
Solve. $\frac{5}{13}x = -1$

$\frac{-13}{5}$

29
Solve. $\frac{9}{5}x = -3$

$\frac{-5}{3}$

30
Solve. $\frac{2}{7}x = \frac{5}{3}$

$\frac{35}{6}$

31
Solve. $\frac{-3}{8}x = \frac{1}{4}$

$\frac{-2}{3}$

32
The reciprocal of 7 is $\frac{1}{7}$. Find the solution for $7x = 10$.

$\frac{10}{7}$

33
The reciprocal of -4 is $\frac{-1}{4}$. Find a solution for $-4x = 7$.

$\frac{-7}{4}$

34
Solve. $-3x = 8$

$\frac{-8}{3}$

35
Solve. $9x = 7$

$\frac{7}{9}$

36
Solve. $-5x = -7$

$\frac{7}{5}$

37
Solve. $4x = 15$

$\frac{15}{4}$

38
Solve. $-5x = 21$

$\frac{-21}{5}$

39
Solve. $-3x = -17$

$\frac{17}{3}$

40
The coefficient of x in $-x = 4$ is -1.
What is the reciprocal of -1?

-1, because $-1 \bullet -1 = 1$

41
To find the solution for $-x = 11$, generate an equivalent equation by multiplying both sides by the reciprocal of the coefficient of x.

$$\begin{aligned} -x &= 11 \\ -1x &= 11 \\ -1 \bullet -1x &= -1 \bullet 11 \\ 1x &= -11 \end{aligned}$$

The solution for $-x = 11$ is ______.

-11

42
Solve. $-x = -7$

7

43
The reciprocal of $\frac{1}{8}$ is 8. Find the solution for $\frac{1}{8}x = 4$.

32

44
Solve. $\frac{-1}{3}x = 5$

-15

45
The equations $\frac{3}{7}x = \frac{5}{8}$ and $x + \frac{3}{7} = \frac{5}{8}$ involve the same numbers, but are the equations equivalent?

No, they have different solutions.

46
The number $\frac{3}{7}$ is a coefficient in $\frac{3}{7}x = \frac{5}{8}$ and is an addend in $x + \frac{3}{7} = \frac{5}{8}$. To solve $\frac{3}{7}x = \frac{5}{8}$, the ________ (opposite, reciprocal) of $\frac{3}{7}$ is used to generate an equivalent equation.

reciprocal

47
The number $\frac{3}{7}$ is a coefficient in $\frac{3}{7}x = \frac{5}{8}$ and is an addend in $x + \frac{3}{7} = \frac{5}{8}$. To solve $x + \frac{3}{7} = \frac{5}{8}$, the ________ (opposite, reciprocal) of $\frac{3}{7}$ is used to generate an equivalent equation.

opposite

48

In $\frac{6}{7}x = 5$, the left side of the equation is a **multiplication expression** and the solution requires the use of the ___________ (opposite, reciprocal) of $\frac{6}{7}$.

reciprocal

49

In $x + \frac{6}{7} = 5$, the left side of the equation is an **addition expression** and the solution requires the use of the ___________ (opposite, reciprocal) of $\frac{6}{7}$.

opposite

50

To solve $\frac{6}{7}x = 5$, both sides are multiplied by $\frac{7}{6}$.

$$\begin{aligned} \frac{6}{7}x &= 5 \\ \frac{7}{6} \bullet \frac{6}{7}x &= \frac{7}{6} \bullet 5 \\ 1x &= \frac{35}{6} \\ x &= \frac{35}{6} \end{aligned}$$

To solve $x + \frac{6}{7} = 5$, should both sides be multiplied by $\frac{7}{6}$?

No

51

To solve $x + \frac{6}{7} = 5$, $\frac{-6}{7}$ should be added to both sides of the equation.

$$\begin{aligned} x + \frac{6}{7} &= 5 \\ x + \frac{6}{7} - \frac{6}{7} &= 5 - \frac{6}{7} \\ x + 0 &= \frac{29}{7} \\ x &= \frac{29}{7} \end{aligned}$$

Are $\frac{6}{7}x = 5$ and $x + \frac{6}{7} = 5$ solved in the same way?

No

52

Which of the following is solved by adding $\frac{-5}{8}$ to both sides of the equation?

$\frac{5}{8}x = \frac{2}{3}$ or $x + \frac{5}{8} = \frac{2}{3}$

$x + \frac{5}{8} = \frac{2}{3}$

53

To solve $\frac{7}{11}x = \frac{1}{2}$, which of the following steps should be taken?

a. Multiply both sides of the equation by $\frac{11}{7}$.

b. Add $\frac{-7}{11}$ to both sides of the equation.

a

54
To solve $x - \frac{2}{3} = \frac{7}{10}$, which of the following steps should be taken?
a. Multiply both sides of the equation by $\frac{-3}{2}$.
b. Add $\frac{2}{3}$ to both sides of the equation.
b

55
To solve $5x = \frac{-7}{8}$, which of the following steps should be taken?
a. Multiply both sides of the equation by $\frac{1}{5}$.
b. Add -5 to both sides of the equation.
a

56
To solve $\frac{-5}{3}x = 7$, which of the following steps should be taken?
a. Multiply both sides of the equation by $\frac{-5}{3}$.
b. Multiply both sides of the equation by $\frac{-3}{5}$.
c. Add $\frac{5}{3}$ to both sides of the equation.
d. Add $\frac{-3}{5}$ to both sides of the equation.
b

57
To solve $x - \frac{3}{8} = \frac{1}{5}$, which of the following steps should be taken?
a. Multiply both sides of the equation by $\frac{-8}{3}$.
b. Multiply both sides of the equation by $\frac{3}{8}$.
c. Add $\frac{3}{8}$ to both sides of the equation.
d. Add $\frac{-8}{3}$ to both sides of the equation.
c

58
To solve $\frac{2}{3}x = \frac{-5}{9}$, which of the following steps should be taken?
a. Multiply both sides of the equation by $\frac{-2}{3}$.
b. Multiply both sides of the equation by $\frac{3}{2}$.
c. Add $\frac{-2}{3}$ to both sides of the equation.
d. Add $\frac{3}{2}$ to both sides of the equation.
b

59
To solve $x + \frac{2}{3} = \frac{-5}{7}$, which of the following steps should be taken?
a. Multiply both sides of the equation by $\frac{-2}{3}$.
b. Multiply both sides of the equation by $\frac{3}{2}$.
c. Add $\frac{-2}{3}$ to both sides of the equation.
d. Add $\frac{3}{2}$ to both sides of the equation.
c

FEEDBACK UNIT 2

This quiz reviews the preceding unit. Answers are at the back of the book.

1. To solve $x + \frac{2}{3} = 5$, which of the following statements is correct?
   a. $\frac{-2}{3}$ should be added to both sides of the equation.
   b. Both sides of the equation should be multiplied by $\frac{3}{2}$.

2. To solve $\frac{-4}{9}x = 7$, which of the following statements is correct?
   a. $\frac{4}{9}$ should be added to both sides of the equation.
   b. Both sides of the equation should be multiplied by $\frac{-9}{4}$.

3. Solve. $5x = 13$

4. Solve. $\frac{-3}{8}x = \frac{1}{2}$

5. Solve. $x - \frac{1}{3} = \frac{1}{2}$

6. Solve. $-x = \frac{-7}{8}$

7. Solve. $\frac{3}{7}x = -2$

UNIT 3: FINDING TRUTH SETS

The following mathematical term is crucial to an understanding of this unit.

Truth set

1

The **truth set** of $5x + 7 = 13$ contains all solutions of the equation.

$$\begin{aligned} 5x + 7 &= 13 \\ 5x + 7 - 7 &= 13 - 7 \\ 5x &= 6 \\ \frac{1}{5} \bullet 5x &= \frac{1}{5} \bullet 6 \\ x &= \frac{6}{5} \end{aligned}$$

Show the truth set of $5x + 7 = 13$.

$\left\{\frac{6}{5}\right\}$

2
The truth set of $9 - 4x = 15$ is found in the following steps.

$$\begin{aligned} 9 - 4x &= 15 \\ 9 - 4x - 9 &= 15 - 9 \\ -4x &= 6 \\ \tfrac{-1}{4} \bullet -4x &= \tfrac{-1}{4} \bullet 6 \\ x &= \tfrac{-3}{2} \end{aligned}$$

Show the truth set of $9 - 4x = 15$.

$\left\{\frac{-3}{2}\right\}$

3
Find the truth set of $2x + 3 = 14$ by adding opposites and multiplying reciprocals.

$2x = 11$
$\left\{\frac{11}{2}\right\}$

4
Find the truth set of $3x + 17 = 12$ by adding opposites and multiplying reciprocals.

$3x = -5$
$\left\{\frac{-5}{3}\right\}$

5
To find the truth set of $4x + 3 = 21$, the following steps are used:

$$\begin{aligned} 4x + 3 &= 21 \\ 4x + 3 - 3 &= 21 - 3 \\ 4x &= 18 \\ \tfrac{1}{4} \bullet 4x &= \tfrac{1}{4} \bullet 18 \\ x &= \tfrac{9}{2} \end{aligned}$$

Show the truth set of $4x + 3 = 21$.

$\left\{\frac{9}{2}\right\}$

6
The first step in finding the truth set for $2x + 7 = 16$ is to add -7 to both sides of the equation. What is to be added to both sides of the equation $3x + 8 = 18$?

-8

7
Find the truth set for $7x - 5 = 19$ by first adding the opposite of -5 to both sides of the equation.

$7x = 24$
$\left\{\frac{24}{7}\right\}$

8
Find the truth set of $-9x - 7 = 4$.

$\left\{\frac{-11}{9}\right\}$

9
Find the truth set of $6x + 9 = 5$.

$\left\{\frac{-2}{3}\right\}$

10
Find the truth set of $2 - 3x = -8$.
[Note: $2 - 3x$ is equivalent to $-3x + 2$.]

$\left\{\frac{10}{3}\right\}$

11
Find the truth set of $11x + 4 = -3$.

$\left\{\frac{-7}{11}\right\}$

12
Find the truth set of $4x - 9 = 14$.

$\left\{\frac{23}{4}\right\}$

13
Find the truth set of $4 - 3x = -4$.

$\left\{\frac{8}{3}\right\}$

14
Find the truth set of $7x + 2 = -8$.

$\left\{\frac{-10}{7}\right\}$

15
Find the truth set of $15x + 4 = 14$.

$\left\{\frac{2}{3}\right\}$

16
Find the truth set of $9 - 8x = 5$.

$\left\{\frac{1}{2}\right\}$

17
Find the truth set of $13 = 7x - 1$. It may be easier to write $7x - 1 = 13$.

$\{2\}$

18
Find the truth set of $15 = 4x - 1$.

$\{4\}$

19
Find the truth set of $23 = 5x + 4$.

$\left\{\frac{19}{5}\right\}$

20
Find the truth set of $4 = 9x + 3$.

$\left\{\frac{1}{9}\right\}$

21
To solve $4(x + 7) = 25$, first remove the parentheses.

$$\begin{aligned} 4(x+7) &= 25 \\ 4x + 28 &= 25 \\ 4x + 28 - 28 &= 25 - 28 \\ 4x &= -3 \\ \frac{1}{4} \bullet 4x &= \frac{1}{4} \bullet -3 \\ x &= \frac{-3}{4} \end{aligned}$$

What is the truth set of $4(x + 7) = 25$?

$\left\{\frac{-3}{4}\right\}$

22
Find the truth set of $3(x + 5) = 7$ by first removing the parentheses.

$3x + 15 = 7$
$\{\frac{-8}{3}\}$

23
Find the truth set of $-2(x + 5) = -7$.

$\{\frac{-3}{2}\}$

24
Find the truth set of $7(2x + 1) = 19$.

$\{\frac{6}{7}\}$

25
Find the truth set of $-2(3x - 4) = 3$.

$\{\frac{5}{6}\}$

26
Find the truth set of $2(3x - 4) = 9$.

$\{\frac{17}{6}\}$

27
Find the truth set of $5(2x - 3) = 16$.

$\{\frac{31}{10}\}$

28
Find the truth set of $15 = 3(5x + 7)$.

$\{\frac{-2}{5}\}$

29
The variable x appears on both sides of the equation $5x + 3 = 2x - 8$. To solve, first add the opposite of 2x to both sides of the equation.

$$\begin{aligned} 5x + 3 &= 2x - 8 \\ 5x + 3 - 2x &= 2x - 8 - 2x \\ 3x + 3 &= -8 \\ 3x + 3 - 3 &= -8 - 3 \\ 3x &= -11 \\ x &= \frac{-11}{3} \end{aligned}$$

Find the truth set of $7x - 3 = 5x + 9$ by first adding $-5x$ to both sides.

$2x - 3 = 9$
$\{6\}$

30
Find the truth set of $4x - 8 = 6 - 3x$ by first adding 3x to both sides.

$7x - 8 = 6$
$\{2\}$

31
Find the truth set of $2x + 9 = 3 + 5x$ by first adding $-5x$ to both sides.

$-3x + 9 = 3$
$\{2\}$

32
Find the truth set of $4x - 3 = 5 - 3x$.

$7x - 3 = 5$
$\left\{\frac{8}{7}\right\}$

33
Find the truth set of $6x - 11 = 2x + 4$.

$4x - 11 = 4$
$\left\{\frac{15}{4}\right\}$

34
Find the truth set of $9x + 3 = 22 + 4x$.

$\left\{\frac{19}{5}\right\}$

35
Find the truth set of $5x - 7 = 6 + x$.

$\left\{\frac{13}{4}\right\}$

36
Find the truth set of $2x + 4 = 6 - x$.

$\left\{\frac{2}{3}\right\}$

37
Find the truth set of $4x - 3 = 2x + 6$.

$\left\{\frac{9}{2}\right\}$

38
Find the truth set of $x + 5 = 3x - 2$.

$\left\{\frac{7}{2}\right\}$

39
Find the truth set of $6x + 3 = x - 7$.

$\{-2\}$

40
To find the truth set of $7 + 3(x + 1) = x + 4$,
first remove the parentheses.

$$\begin{aligned} 7 + 3(x + 1) &= x + 4 \\ 7 + 3x + 3 &= x + 4 \\ 3x + 10 &= x + 4 \end{aligned}$$

Complete finding the truth set of $7 + 3(x + 1) = x + 4$.

$\{-3\}$

41
To find the truth set of $2x + 3(x + 2) = 2(x - 5) + 8$,
first simplify each side of the equation separately.
The two simplifications are shown below.

$$\begin{aligned} 2x + 3(x + 2) &= 2(x - 5) + 8 \\ 2x + 3x + 6 &= 2x - 10 + 8 \\ 5x + 6 &= 2x - 2 \end{aligned}$$

Complete finding the truth set of $2x + 3(x + 2) = 2(x - 5) + 8$.

$\left\{\frac{-8}{3}\right\}$

42
Find the truth set of $4x - 3(x + 5) = 5 - 2x$ by first
simplifying the left side of the equation.

$\left\{\frac{20}{3}\right\}$

43
Find the truth set of $3(2x - 4) + 8 = 5 - (x + 2)$
by first simplifying each side of the equation. {1}

44
Find the truth set of $3(5x - 4) - 8 = 2(3x - 1)$. {2}

45
Find the truth set of $(4x - 7) = 2(x - 3) + 5$. {3}

46
Find the truth set of $2(3x + 4) + 4 = 3(x - 5)$. {-9}

FEEDBACK UNIT 3

This quiz reviews the preceding unit. Answers are at the back of the book.

Find the truth set for each of the following equations.

1. $x + \frac{3}{4} = \frac{4}{5}$
2. $x - \frac{7}{8} = \frac{-3}{5}$
3. $\frac{-4}{5}x = \frac{3}{7}$
4. $\frac{9}{10}x = \frac{-6}{7}$
5. $-5x = 14$
6. $-3x = -2$
7. $5x - 4 = 17$
8. $-9x + 4 = 8$
9. $4 = 7x - 5$
10. $3(x + 4) = 9$
11. $2x + 3(x + 4) = 4(x - 2) + 9$

UNIT 4: SOLVING AND CHECKING FRACTIONAL EQUATIONS

1
The counting number multiples of 3 are: 3, 6, 9, 12, . . .
The counting number multiples of 7 are: ____________ 7, 14, 21, 28, . . .

2
The **common multiples** of 3 and 7 are: 21, 42, 63, 84, . . .
The common multiples of 4 and 5 are: ______________ 20, 40, 60, 80, . . .

3
The common multiples of 6 and 4 are: 12, 24, 36, 48, . . .
The common multiples of 6 and 9 are: ________________

18, 36, 54, 72, . . .

4
The **least common multiple** (LCM) of 3 and 7 is 21 because it is the smallest counting number that is a common multiple. What is the LCM of 4 and 5?

20

5
What is the LCM of 6 and 4?

12

6
What is the LCM of 6 and 9?

18

7
One method for solving $\frac{5}{12}x - \frac{3}{8} = \frac{1}{4}$ is shown below. It involves adding opposites and multiplying by reciprocals.

$$\frac{5}{12}x - \frac{3}{8} = \frac{1}{4}$$

a. Add $\frac{3}{8}$ to both sides. $\quad \frac{5}{12}x - \frac{3}{8} + \frac{3}{8} = \frac{1}{4} + \frac{3}{8}$

$$\frac{5}{12}x + 0 = \frac{5}{8}$$

b. Multiply $\frac{12}{5}$ on both sides. $\quad \frac{12}{5} \bullet \frac{5}{12}x = \frac{12}{5} \bullet \frac{5}{8}$

$$x = \frac{3}{2}$$

Another method for solving the same equation depends upon finding the LCM of the denominators. What is the LCM of the denominators?

LCM of 12, 8 and 4 is 24.

8
To solve $\frac{5}{12}x - \frac{3}{8} = \frac{1}{4}$ with a minimum amount of arithmetic, first multiply each term by the LCM of the denominators.

$$\frac{5}{12}x - \frac{3}{8} = \frac{1}{4}$$

a. Multiply by LCM $\quad 24 \bullet \frac{5}{12}x - 24 \bullet \frac{3}{8} = 24 \bullet \frac{1}{4}$

b. Cancel where possible $\quad 2 \bullet \frac{5}{1}x - 3 \bullet \frac{3}{1} = 6 \bullet \frac{1}{1}$

c. Simplify $\quad 10x - 9 = 6$

d. Complete the solution $\quad 10x = 15$

$$x = \frac{3}{2}$$

The solution for $\frac{5}{12}x - \frac{3}{8} = \frac{1}{4}$ is _______.

$\frac{3}{2}$

9

The equations $\frac{5}{12}x - \frac{3}{8} = \frac{1}{4}$ and $10x - 9 = 6$ are equivalent. Which is easier to solve?

$10x - 9 = 6$

10

The equation $10x - 9 = 6$ is relatively easy to solve because its numbers, 10, -9, and 6, are integers. If $\frac{3}{2}x - \frac{5}{6} = \frac{1}{4}$ and $18x - 10 = 3$ are equivalent equations, which would be easier to solve?

$18x - 10 = 3$

11

To solve $\frac{3}{2}x - \frac{5}{6} = \frac{1}{4}$, first multiply each term by the LCM of the denominators.

$$\begin{aligned} \tfrac{3}{2}x - \tfrac{5}{6} &= \tfrac{1}{4} \\ 12 \bullet \tfrac{3}{2}x - 12 \bullet \tfrac{5}{6} &= 12 \bullet \tfrac{1}{4} \\ 6 \bullet \tfrac{3}{1}x - 2 \bullet \tfrac{5}{1} &= 3 \bullet \tfrac{1}{1} \\ 18x - 10 &= 3 \end{aligned}$$

Are $\frac{3}{2}x - \frac{5}{6} = \frac{1}{4}$ and $18x - 10 = 3$ equivalent equations?

Yes

12

Since $\frac{3}{2}x - \frac{5}{6} = \frac{1}{4}$ and $18x - 10 = 3$ are equivalent equations, solve one of them to find the solution for both.

$18x = 13$, $x = \frac{13}{18}$

13

To solve $\frac{2}{3}x - \frac{4}{9} = \frac{5}{6}$, first multiply each term by the common denominator, 18.

$$\begin{aligned} \tfrac{2}{3}x - \tfrac{4}{9} &= \tfrac{5}{6} \\ 18 \bullet \tfrac{2}{3}x - 18 \bullet \tfrac{4}{9} &= 18 \bullet \tfrac{5}{6} \\ 6 \bullet \tfrac{2}{1}x - 2 \bullet \tfrac{4}{1} &= 3 \bullet \tfrac{5}{1} \\ 12x - 8 &= 15 \end{aligned}$$

Find the solution for $\frac{2}{3}x - \frac{4}{9} = \frac{5}{6}$.

$12x = 23$, $x = \frac{23}{12}$

14

To eliminate the denominators (change them to 1's) of $\frac{1}{3}x + \frac{3}{4} = \frac{1}{6}$, each term is multiplied by the common denominator, 12.

$$\frac{1}{3}x + \frac{3}{4} = \frac{1}{6}$$
$$12 \cdot \frac{1}{3}x + 12 \cdot \frac{3}{4} = 12 \cdot \frac{1}{6}$$
$$4 \cdot \frac{1}{1}x + 3 \cdot \frac{3}{1} = 2 \cdot \frac{1}{1}$$
$$4x + 9 = 2$$

Is $4x + 9 = 2$ equivalent to $\frac{1}{3}x + \frac{3}{4} = \frac{1}{6}$?

Yes

15

Since $\frac{1}{3}x + \frac{3}{4} = \frac{1}{6}$ and $4x + 9 = 2$ are equivalent equations, solve one of them to find the solution for both.

$4x = -7$, $x = \frac{-7}{4}$

16

The denominators of a fractional equation can be changed to 1's by multiplying each term of the equation by the common denominator of its fractions. What is the common denominator for $\frac{1}{5}x - \frac{2}{15} = \frac{7}{10}$?

30

17

Multiply each term of $\frac{1}{5}x - \frac{2}{15} = \frac{7}{10}$ by the common denominator of the fractions. After simplifying, what new, equivalent equation is obtained?

$6x - 4 = 21$

18

Since $\frac{1}{5}x - \frac{2}{15} = \frac{7}{10}$ and $6x - 4 = 21$ are equivalent equations, solve one of the equations and find the solution for both.

$\frac{25}{6}$

19

Complete the solution for $\frac{2}{3}x - 4 = \frac{1}{5}$.

$$\begin{aligned} \frac{2}{3}x - 4 &= \frac{1}{5} \\ 15 \cdot \frac{2}{3}x - 15 \cdot 4 &= 15 \cdot \frac{1}{5} \\ 5 \cdot \frac{2}{1}x - 15 \cdot 4 &= 3 \cdot \frac{1}{1} \\ 10x - 60 &= 3 \end{aligned}$$

____

$\frac{63}{10}$

20

Show the first step in solving $2x - \frac{3}{4} = \frac{7}{10}$.

$20 \cdot 2x - 20 \cdot \frac{3}{4} = 20 \cdot \frac{7}{10}$

21

Complete the solution for $2x - \frac{3}{4} = \frac{7}{10}$.

$$\begin{aligned} 20 \cdot 2x - 20 \cdot \frac{3}{4} &= 20 \cdot \frac{7}{10} \\ 40x - 15 &= 14 \end{aligned}$$

____

$\frac{29}{40}$

22

Show the first step in solving $\frac{-3}{4}x + \frac{5}{6} = \frac{-7}{12}$.

$12 \cdot \frac{-3}{4}x + 12 \cdot \frac{5}{6} = 12 \cdot \frac{-7}{12}$

23

Complete the solution for $\frac{-3}{4}x + \frac{5}{6} = \frac{-7}{12}$.

$$\begin{aligned} 12 \cdot \frac{-3}{4}x + 12 \cdot \frac{5}{6} &= 12 \cdot \frac{-7}{12} \\ -9x + 10 &= -7 \end{aligned}$$

____

$\frac{17}{9}$

24

Solve. $6x - \frac{3}{10} = \frac{4}{5}$

$\frac{11}{60}$

25

Solve. $\frac{5}{12}x - \frac{3}{4} = \frac{7}{8}$

$\frac{39}{10}$

26

Solve. $\frac{5}{6}x - 9 = \frac{2}{3}$

$\frac{58}{5}$

27

Solve. $\frac{-3}{8}x + \frac{1}{4} = \frac{2}{3}$

$\frac{-10}{9}$

28

Solve. $\frac{2}{3}x - \frac{1}{6} = \frac{1}{2}x + \frac{11}{12}$

$\frac{13}{2}$

29

Solve. $x - \frac{7}{15} = \frac{3}{5}x + \frac{9}{10}$

$\frac{41}{12}$

30

Solve. $\frac{2}{7}x - \frac{11}{14} = \frac{1}{3}x + \frac{5}{6}$

-34

31

Solve. $\frac{1}{3}x - \frac{3}{4} = \frac{1}{2}$

$\frac{15}{4}$

32

To check $\frac{15}{4}$ as a solution for $\frac{1}{3}x - \frac{3}{4} = \frac{1}{2}$, replace x in the equation and show that the statement is true.

$$\frac{1}{3}x - \frac{3}{4} = \frac{1}{2}$$
$$\frac{1}{3} \cdot \frac{15}{4} - \frac{3}{4} = \frac{1}{2}$$
$$\frac{5}{4} - \frac{3}{4} = \frac{1}{2}$$

Is $\frac{5}{4} - \frac{3}{4} = \frac{1}{2}$ a true statement?

$\frac{1}{2} = \frac{1}{2}$ is true.

33

The solution for $\frac{1}{3}x - \frac{3}{4} = \frac{1}{2}$ is $\frac{15}{4}$ because when x is replaced by $\frac{15}{4}$, the equation becomes the true statement $\frac{1}{2} = \frac{1}{2}$. Replace x by 3 in $\frac{2}{5}x - \frac{1}{2} = \frac{3}{10}$ and determine whether it is a solution.

$\frac{6}{5} - \frac{1}{2} = \frac{3}{10}$

$\frac{7}{10} = \frac{3}{10}$ is false.

3 is not a solution.

34

Solve and check $\frac{2}{5}x - \frac{1}{2} = \frac{3}{10}$.

$x = 2$

$\frac{4}{5} - \frac{1}{2} = \frac{3}{10}$

$\frac{8}{10} - \frac{5}{10} = \frac{3}{10}$

$\frac{3}{10} = \frac{3}{10}$ is true.

35

Determine if $\frac{3}{5}$ is a solution for $\frac{5}{6}x - \frac{2}{9} = \frac{2}{3}$ by evaluating the numerical expression on the left side of the equality below.

$$\frac{5}{6}x - \frac{2}{9} = \frac{2}{3}$$

If $x = \frac{3}{5}$ $\quad \frac{5}{6} \bullet \frac{3}{5} - \frac{2}{9} = \frac{2}{3}$

$$\_\_\_\_\_\_ = \frac{2}{3}$$

$\frac{1}{2} - \frac{2}{9} = \frac{2}{3}$

$\frac{9}{18} - \frac{4}{18} = \frac{2}{3}$

$\frac{5}{18} = \frac{2}{3}$ is false.

$\frac{3}{5}$ is not a solution.

36

Solve and check. $\frac{5}{6}x - \frac{2}{9} = \frac{2}{3}$

$x = \frac{16}{15}$

$\frac{5}{6} \bullet \frac{16}{15} - \frac{2}{9} = \frac{2}{3}$

$\frac{8}{9} - \frac{2}{9} = \frac{2}{3}$

$\frac{6}{9} = \frac{2}{3}$ is true.

37

To determine if $\frac{2}{5}$ is a solution for $\frac{-3}{10}x + \frac{7}{15} = \frac{3}{5} - \frac{19}{30}x$, each side of the equality is separately evaluated. Evaluate the numerical expression on the right side of the equality below and state whether $\frac{2}{5}$ is a solution.

$$\frac{-3}{10}x + \frac{7}{15} = \frac{3}{5} - \frac{19}{30}x$$

If $x = \frac{2}{5}$ $\quad \frac{-3}{10} \bullet \frac{2}{5} + \frac{7}{15} = \frac{3}{5} - \frac{19}{30} \bullet \frac{2}{5}$

$$\frac{-3}{25} + \frac{7}{15} = \_\_\_\_\_\_$$

$$\frac{-9}{75} + \frac{35}{75} = \_\_\_\_\_\_$$

$$\frac{26}{75} = \_\_\_\_\_\_$$

$\frac{3}{5} - \frac{19}{75}$

$\frac{45}{75} - \frac{19}{75}$

$\frac{26}{75}$

$\frac{26}{75} = \frac{26}{75}$ is true.

$\frac{2}{5}$ is a solution.

38

Solve and check. $\frac{2}{3}x - \frac{7}{10} = x - \frac{1}{6}$

$x = \frac{-8}{5}$

Check: $\frac{-53}{30} = \frac{-53}{30}$ is true.

39

Solve and check. $3x - \frac{4}{5} = \frac{7}{10}$

$x = \frac{1}{2}$

Check: $\frac{7}{10} = \frac{7}{10}$ is true.

40

Solve and check. $\frac{3}{8}x - \frac{2}{3} = x - \frac{3}{4}$

$x = \frac{2}{15}$

Check: $\frac{-37}{60} = \frac{-37}{60}$ is true.

41

Solve and check. $\frac{-5}{16} - 4x = \frac{3}{8}$

$x = \frac{-11}{64}$

Check: $\frac{3}{8} = \frac{3}{8}$ is true.

42

Solve and check. $\frac{5}{8}x - \frac{1}{2} = \frac{3}{4}$

$x = 2$

Check: $\frac{3}{4} = \frac{3}{4}$ is true.

43

Solve and check. $\frac{7}{12}x - \frac{3}{8} = \frac{5}{6}$

$x = \frac{29}{14}$

Check: $\frac{5}{6} = \frac{5}{6}$ is true.

FEEDBACK UNIT 4

This quiz reviews the preceding unit. Answers are at the back of the book.

Solve and check each of the following equations.

1. $\frac{-3}{4}x - \frac{7}{8} = \frac{5}{6}$
2. $\frac{4}{5}x - 2 = \frac{3}{10}$
3. $\frac{1}{2}x - \frac{2}{3} = \frac{11}{12}x + \frac{1}{6}$
4. $\frac{4}{5}x - \frac{1}{2} = \frac{7}{10} - x$
5. $\frac{3}{5} - 2x = \frac{1}{10}$
6. $\frac{7}{8}x - \frac{1}{4} = \frac{-3}{8}$
7. $\frac{3}{4}x + \frac{5}{6} = \frac{1}{3}x + \frac{7}{12}$
8. $-2x + \frac{2}{7} = \frac{5}{14}x - \frac{1}{2}$

UNIT 5: APPLICATIONS

In this Applications Section, the format of the text has been altered. Answers for the problems appear beneath them rather than in the right-hand column. Your studying emphasis should be on learning the best procedures to follow with word problems.

1

To find the missing number described by: A number increased by 5 has a sum of 41, the following steps are used.

a. First translate the sentence into an equation.
A number increased by 5 has a sum of 41 translates as $N + 5 = 41$

b. Solve the equation.
If $N + 5 = 41$, then $N = 36$

c. Check 36 in the original sentence to see if it really meets the description.
Yes, 36 increased by 5 does have a sum of 41.

Find the missing number described by: A number increased by 3 gives 19.

Answer: a. $N + 3 = 19$
b. $N = 16$
c. Yes, 16 increased by 3 does give 19.

2

To find the missing number described by: A number decreased by 8 is 23, the following steps are used.

a. First translate the sentence into an equation.
   A number decreased by 8 is 23 translates as $N - 8 = 23$

b. Solve the equation.
   If $N - 8 = 23$, then $N = 31$

c. Check 31 in the original sentence to see if it really meets the description.
   Yes, 31 decreased by 8 is 23.

Find the missing numbers described by:
9 less than a number is 16.

Answer:
a. $N - 9 = 16$
b. $N = 25$
c. Yes, 9 less than 25 is 16.

3

To find the missing number described by: The product of a number and 8 is 24, the following steps are used.

a. First translate the sentence into an equation.
   The product of a number and 8 is 24 translates as $8N = 24$

b. Solve the equation.
   If $8N = 24$, then $N = 3$

c. Check 3 in the original sentence to see if it really meets the description.
   Yes, the product of 3 and 8 is 24.

Find the missing numbers described by:
A number increased threefold is 45.

Answer:
a. $3N = 45$
b. $N = 15$
c. Yes, 15 increased threefold is 45.

4

To find the missing number described by: Twice a number plus 7 is 23, the following steps are used.

a. First translate the sentence into an equation.
   Twice a number plus 7 is 23 translates as $2N + 7 = 23$

b. Solve the equation.
   If $2N + 7 = 23$, then $2N = 16$ and $N = 8$

c. Check 8 in the original sentence to see if it really meets the description.
   Yes, twice 8 plus 7 is 23.

Find the missing numbers described by: 13 plus 5 times a number gives 28.

Answer: a. $5N + 13 = 28$
b. $N = 3$
c. Yes, 13 plus 5 times 3 gives 28.

5

To find the missing number described by: Decreasing 5 times a number by 4 gives 31, the following steps are used.

a. First translate the sentence into an equation.
   Decreasing 5 times a number by 4 gives 31 translates as $5N - 4 = 31$

b. Solve the equation.
   If $5N - 4 = 31$, then $5N = 35$ and $N = 7$

c. Check 7 in the original sentence to see if it really meets the description.
   Yes, decreasing 5 times 7 by 4 does give 31.

Find the missing numbers described by: 6 less than the product of a number and 7 is 22.

Answer: a. $7N - 6 = 22$
b. $N = 4$
c. Yes, 6 less than the product of 4 and 7 is 22.

FEEDBACK UNIT 5 FOR APPLICATIONS

1. Find the number described by: A number increased by 11 equals 48.

2. Find the number described by: 15 less than a number is 17.

3. Find the number described by: 7 multiplied by a number gives 56.

4. Find the number described by: 19 more than twice a number equals 27.

5. Find the number described by: 4 less than the product of a number and 9 is 59.

SUMMARY FOR SOLVING EQUATIONS WITH THE RATIONAL NUMBERS

The following mathematical terms are crucial to an understanding of this chapter.

Equivalent equations
Solve
Truth set

In Chapter 9, methods for finding the truth set using the set of rational numbers were presented. The method that was stressed for solving fractional equations was to multiply both sides of the equation by the least common denominator to eliminate all fractions from the equation.

Unit 4 presented a method for checking the solution of fractional equations that had rational solutions.

CHAPTER 9 MASTERY TEST

The following questions test the objectives of Chapter 9. Answers are at the back of the book. The number in parentheses which follows each problem indicates the unit in which it can be learned.

For problems 1-17, solve each equation.

1. $5x + 3 = 2$ (3)
2. $4x - 7 = 16$ (3)
3. $-6x + 12 = 36$ (3)
4. $2x - 5 = 3x + 7$ (3)
5. $x + 7 = 3x - 54$ (3)
6. $4 - 6x = 17 - x$ (3)
7. $5 + 4x = 9 - 2x$ (3)
8. $x + \frac{2}{3} = \frac{5}{6}$ (1)
9. $x - \frac{5}{8} = \frac{1}{4}$ (1)
10. $\frac{5}{3}x + 1 = \frac{2}{7}$ (3)
11. $\frac{-5}{2}x = \frac{1}{5}$ (2)
12. $4(2x + 3) - 6 = x + 1$ (3)
13. $5 - (6x + 7) = 7$ (3)
14. $3(2x - 1) - 4 = -(x + 5) + 3$ (3)
15. $\frac{-2}{3}x - \frac{5}{6} = \frac{3}{8}$ (3)
16. $\frac{3}{4}x - \frac{1}{5} = x - \frac{3}{10}$ (3)
17. $\frac{-2}{7}x + 1 = \frac{2}{3}$ (3)

For problems 18-20, solve and check each equation.

18. $\frac{3}{5}x - \frac{1}{2} = \frac{-3}{10}$ (4)
19. $\frac{1}{4} + 3x = \frac{-7}{12} - \frac{5}{6}x$ (4)
20. $\frac{1}{6}x + \frac{3}{5} = 1 - \frac{3}{10}x$ (4)
21. Find the number described by: 17 plus 6 times a number gives 65. (5)
22. Find the number described by: 9 less than the product of a number and 6 is 15. (5)

Answers for All Tests and Feedback Exercises

Chapter 1
Objectives

1. {12, 14, 16, 18}
2. {19, 21, 23}
3. { }
4. false
5. true
6. true
7. 16
8. 15
9. 72
10. 7
11. 34
12. 32
13. 27
14. 19
15. {1, 3, 5, 15}
16. 8
17. 30

Chapter 1
Feedback Unit 1

1. set
2. {2, 4, 8, 9}
3. {8, 9}
4. { }
5. {8, 10}
6. {7, 9, 11}
7. { }
8. no
9. yes
10. no
11. a. $\in$ b. $\notin$
    c. $\notin$ d. $\in$
    e. $\notin$

12. true
13. false
14. true
15. false
16. false
17. false
18. true
19. true

Chapter 1
Feedback Unit 2

1. two
2. yes
3. yes
4. yes
5. yes
6. yes
7. yes
8. yes

Chapter 1
Feedback Unit 3

1. yes
2. no
3. yes
4. no
5. 8
6. 12
7. 6
8. 14

Chapter 1
Feedback Unit 4

1. two
2. 48
3. 140
4. yes, 42
5. yes, 54
6. 96
7. 120
8. 48

Chapter 1
Feedback Unit 5

1. yes
2. no
3. yes
4. no
5. 1
6. 3
7. 6
8. 4

Chapter 1
Feedback Unit 6

1. 8
2. 36
3. 9
4. 19
5. 102
6. 26
7. 15
8. 39
9. 21

10. 88
11. 50
12. 38

Chapter 1 Feedback Unit 7

1. 10
2. 3
3. 1 • 10, 2 • 5
4. {1, 2, 7, 14}
5. {1, 5, 25}
6. {1, 2, 4, 8, 16}
7. {1, 2, 3, 5, 6, 10, 15, 30}
8. {1, 2, 4, 8, 16, 32}
9. {1, 31}
10. 6
11. 10
12. 1
13. 4
14. 6
15. 8
16. 48
17. 28
18. 60
19. 105
20. 42

Chapter 1 Feedback Unit 8

1. 17 pounds
2. 268 pounds
3. 116 inches
4. 187 sq feet
5. $357

Chapter 1 Mastery Test

1. {4, 7, 15}
2. {12, 14, 16}
3. {15, 17, 19}
4. { }
5. { }
6. true
7. false
8. true
9. true
10. false
11. 21
12. 19
13. 5
14. 3
15. 135
16. 105
17. 2
18. 8
19. 29
20. 21
21. 36
22. 32
23. 90
24. 52
25. {1, 2, 4, 5, 10, 20}
26. 6
27. 3
28. 1
29. 18
30. 563 acres
31. $56,488

Chapter 2 Objectives

1. a, c
2. a, c, d
3. 26
4. 56
5. 22
6. 19 + 7x
7. 9x + (8 + 13)
8. 7x
9. (8 • 9)y
10. 1x
11. (8 + 13)x
12. 12 + 17x
13. 53x + 45
14. 78x + 34

Chapter 2 Feedback Unit 1

1. b, c
2. c, d
3. a. 18
   b. 90
   c. 99

Chapter 2 Feedback Unit 2

1. 8
2. 18
3. 35
4. 21
5. 22
6. 19
7. 11
8. 20
9. 33
10. 27

Chapter 2
Feedback Unit 3

1. a. y + x
   b. a + 13
   c. 8 + w
   d. 7 + 2k
2. a. x + (y + z)
   b. 5 + (x + 9)
   c. 3x + (6 + 7)
   d. (5 + 13) + 6x
3. 5x + 39
4. 7x + 19
5. 8x + 23
6. x + 16
7. x + 12
8. 8 + y or y + 8
9. x + 19
10. 3x + 11

Chapter 2
Feedback Unit 4

1. a. 18x
   b. 9(6x)
2. a. (6 • 3)x
   b. 4(x • 5)
3. a. k
   b. 1 • m
   c. 3x + 2
   d. 14x
4. 35x
5. 8x + 31
6. 15x
7. 12x
8. 10x
9. 21x
10. 32x

Chapter 2
Feedback Unit 5

1. 13x + 6
2. 14x + 7
3. 20x + 9
4. 7x + 9
5. 7x + 5
6. 9x + 13
7. 9x + 18
8. 13x + 7

Chapter 2
Feedback Unit 6

1. 6x + 24
2. 12x + 21
3. 2x + 9
4. 5x + 30
5. 6x + 15
6. 5x + 15
7. 15x + 25
8. 11x + 5
9. 47x + 11
10. 11x + 21

Chapter 2
Feedback Unit 7

1. 25.388 milligrams
2. $768.56
3. 11.72 inches
4. 81.5
5. 250 liters

Chapter 2
Mastery Test

1. 26
2. 72
3. 27
4. 24
5. x + 15
6. 35x
7. 16y + 6
8. 21 + 14x
9. 14x + 5
10. 13x + 34
11. 7x + 10
12. 23x + 24
13. 24x + 23
14. 19x + 4
15. 20x + 6
16. 30x + 75
17. 40x + 110
18. 23x + 37
19. 87x + 103
20. 17x + 22
21. 192 pounds
22. $360

Chapter 3
Objectives

1. 12
2. 7
3. 6
4. 8
5. 9
6. 4
7. 7
8. 6
9. 3
10. 2
11. {6}
12. { }

13. {4}
14. { }
15. {5}
16. { }
17. {1, 2, 3, . . . }
18. 12; 31 = 31 is true
19. 3; 42 = 42 is true
20. 5; 45 = 45 is true

CHAPTER 3 FEEDBACK UNIT 1

1. a, c, d
2. b, c, d
3. 78
4. 8
5. 14
6. 29
7. 3
8. 14
9. 3
10. 443

CHAPTER 3 FEEDBACK UNIT 2

1. 7
2. 8
3. 8
4. 3
5. 3
6. 2
7. 6

CHAPTER 3 FEEDBACK UNIT 3

1. {6}
2. {11}
3. {1, 2, 3, . . . }
4. {14}
5. {1, 2, 3, . . . }
6. {9}
7. { }
8. { }

CHAPTER 3 FEEDBACK UNIT 4

1. 11; 14 = 14 is true
2. 8; 56 = 56 is true
3. 9; 14 = 14 is true
4. 3; 39 = 39 is true
5. 9; 18 = 18 is true

CHAPTER 3 FEEDBACK UNIT 5

1. $71,052.63
2. 40%
3. 30 students
4. 234 girls
5. $120,000

CHAPTER 3 MASTERY TEST

1. 18
2. 8
3. 15
4. 9
5. 11
6. 4
7. 6
8. 5
9. 3
10. 5
11. 6; 51 = 51 is true
12. 4; 60 = 60 is true
13. 1; 19 = 19 is true
14. {7}
15. {14}
16. {1, 2, 3, . . .}
17. { }
18. { }
19. {1, 2, 3, . . .}
20. {3}
21. x = 2; 2 + 11 = 13 is true
22. x = 6; 7 • 6 = 42 is true
23. 3 liters
24. .312 liters

CHAPTER 4 OBJECTIVES

1. 13
2. 5
3. -19
4. 9
5. -12
6. -72
7. 63
8. -42
9. 32
10. -19
11. -6
12. 5
13. 0
14. 84
15. 36
16. 16
17. -16
18. 81
19. 4, 6, 8, 9, 10
20. 17, 19

Chapter 4
Feedback Unit 1

1. integers
2. 5
3. -13
4. 0
5. 0
6. 0
7. 0
8. negative
9. {8}
10. {6}

Chapter 4
Feedback Unit 2

1. 20
2. positive
3. -13
4. negative
5. 4
6. 0
7. -15
8. -8
9. -23
10. 7
11. 0
12. 5
13. -7
14. -14
15. -2
16. -22
17. -12
18. -10
19. 5
20. -11

Chapter 4
Feedback Unit 3

1. 24
2. positive
3. -20
4. negative
5. 21
6. positive
7. 54
8. 0
9. -28
10. 36
11. -4
12. 0
13. -60
14. -8
15. 120
16. -24
17. -20
18. 30
19. -18
20. -30

Chapter 4
Feedback Unit 4

1. -12
2. 9
3. 11
4. -6
5. 3
6. 4
7. -5
8. 2
9. -7
10. -7
11. -16
12. 19
13. -26
14. -46
15. 9

Chapter 4
Feedback Unit 5

1. 60
2. -72
3. 35
4. 13
5. 16
6. -12
7. 69
8. 81
9. -8
10. 1
11. 25
12. -25

Chapter 4
Feedback Unit 6

1. 2, 7, 17, 31
2. 4, 10, 25, 30
3. 2, 3, 5, 7, 11
4. 53, 59
5. 17, 19, 23
6. 32, 33, 34, 35, 36
7. 59, 61
8. 83, 89

Chapter 4
Feedback Unit 7

1. $8,000
2. 24 feet
3. $789.75
4. 5 hours
5. 12.8%

Chapter 4 Mastery Test

1. -63
2. -8
3. -13
4. 9
5. -15
6. -6
7. -30
8. 24
9. -17
10. -41
11. -19
12. 6
13. -6
14. 40
15. -24
16. -125
17. 81
18. 32
19. 19, 23
20. 9, 10, 12, 14
21. $12,200
22. 63 miles per hour

Chapter 5 Objectives

1. 20
2. -8
3. 81
4. (9 + 3) + x
5. 3 + 7x
6. -5x + 4
7. 6x + 3
8. -35x
9. -30x
10. x^6
11. x^2
12. x^6
13. x • 7
14. (-3 • 5) • 9
15. 2
16. -24 + 9x
17. -9x – 1
18. x + 2
19. 5x – 6
20. 35x – 54

Chapter 5 Feedback Unit 1

1. 3
2. 8
3. -13
4. 22
5. 19
6. -125
7. 81
8. 64

Chapter 5 Feedback Unit 2

1. 3w + 8
2. 4y + 13x
3. (6x – 8x) + 3
4. 9x + (-13 + 7)
5. -9x – 7
6. -11x – 14
7. 3x – 5
8. -4x – 7

Chapter 5 Feedback Unit 3

1. 5 • (-7x)
2. -8 • x
3. (-6 • 5)x
4. -4(x • x)
5. 35x
6. -18x
7. 27x
8. -14x

Chapter 5 Feedback Unit 4

1. a. 1x
   b. -1y
   c. 6x – 1(5x – 3)
   d. -3 + 1(4x + 7)
   e. $-1x^4$
2. a. 4x
   b. -10x
   c. -5x – 7
   d. -6x – 8
   e. 8x – 8

Chapter 5 Feedback Unit 5

1. 21x – 35
2. -4x + 12
3. 6x – 9
4. -5x + 5
5. 10x – 3
6. x – 12
7. 8x + 3

CHAPTER 5
FEEDBACK UNIT 6

1. x^5
2. y • y • y • y
3. x^9
4. y^{22}
5. z^{12}
6. x^7
7. x^9
8. x^2

CHAPTER 5
FEEDBACK UNIT 7

1. n + 14 or 14 + n
2. n – 15
3. 7n or 7 • n or n • 7
4. n + 19 or 19 + n
5. n – 11

CHAPTER 5
MASTERY TEST

1. 19
2. -11
3. -243
4. 3 + 4x
5. (9 + 7x) + 5
6. -3x + 17
7. -3
8. -12x
9. -27x
10. x^8
11. x^7
12. x^{10}
13. (4 • 3) • 5
14. x • -2
15. -5x – 10
16. -15x + 35
17. x + 3
18. x – 10
19. -3x – 6
20. -12
21. n – 6
22. n + 8 or 8 + n

CHAPTER 6
OBJECTIVES

1. -3
2. 32
3. -8
4. 8
5. 6
6. 7
7. 3
8. 3
9. 3
10. 6
11. {7}; 25 = 25 is true
12. {-6}; -26 = -26 is true
13. {3}; 5 = 5 is true
14. {4}; -3 = -3 is true

CHAPTER 6
FEEDBACK UNIT 1

1. 15
2. 24
3. -21
4. -35
5. 7
6. -5
7. 9
8. 10
9. -8
10. -19

CHAPTER 6
FEEDBACK UNIT 2

1. -7
2. -4
3. 7
4. -4
5. -2
6. 3
7. -5
8. 5
9. 9
10. -9
11. 11
12. 8

CHAPTER 6
FEEDBACK UNIT 3

1. -10
2. 3
3. -8
4. -4
5. 5
6. 8
7. -3
8. -3
9. -1
10. 20
11. -5
12. 1

CHAPTER 6
FEEDBACK UNIT 4

1. {4}; 17 = 17 is true
2. {-9}; -29 = -29 is true
3. {2}; -3 = -3 is true
4. {3}; 14 = 14 is true
5. {-5}; -11 = -11 is true

Chapter 6 Feedback Unit 5

1. n + 11 = 48
2. n – 15 = 17
3. 7n = 56
4. 2n + 19 = 27
5. 9n – 4 = 67

Chapter 6 Mastery Test

1. -6
2. 18
3. -9
4. 2
5. 5
6. 2
7. 1
8. 2
9. -15
10. 15
11. 7
12. -44
13. 5
14. -2
15. 3
16. 7
17. {3}; 16 = 16 is true
18. {-4}; -35 = -35 is true
19. {0}; -21 = -21 is true
20. {1}; 5 = 5 is true
21. 6n + 17 = 65
22. 6n – 9 = 15

Chapter 7 Objectives

1. $\frac{-3}{0}$
2. zero
3. yes
4. yes
5. $\frac{-3}{4}$
6. $\frac{2}{3}$
7. $\frac{-21}{40}$
8. 0
9. $\frac{6}{35}$
10. 1
11. $\frac{-162}{125}$
12. 8
13. 40
14. 9
15. $\frac{-17}{10}$
16. $\frac{14}{25}$
17. $\frac{23}{60}$
18. $\frac{-1}{5}$
19. $\frac{-14}{9}$
20. -6

Chapter 7 Feedback Unit 1

1. a. $\frac{5}{13}$
   b. $\frac{15}{-7}$
   c. $\frac{-12}{-4}$
   d. $\frac{-9}{2}$
   e. $\frac{0}{5}$
2. no
3. yes
4. yes
5. 15
6. -7
7. yes
8. yes
9. yes
10. no

Chapter 7 Feedback Unit 2

1. $\frac{2}{1}$
2. $\frac{-15}{1}$
3. yes
4. yes
5. no
6. yes
7. no
8. yes
9. no
10. yes
11. $\frac{13}{15}$
12. $\frac{-4}{11}$

Chapter 7 Feedback Unit 3

1. $\frac{3}{5}$
2. $\frac{1}{4}$
3. $\frac{-9}{10}$
4. $\frac{2}{5}$
5. $\frac{-1}{2}$
6. $\frac{-5}{2}$

7. $\frac{-2}{5}$
8. $\frac{4}{7}$

Chapter 7 Feedback Unit 4

1. $\frac{-24}{35}$
2. $\frac{6}{35}$
3. $\frac{-45}{56}$
4. $\frac{-7}{5}$
5. $\frac{3}{7}$
6. $\frac{6}{55}$
7. $\frac{-8}{3}$
8. 1
9. 1

Chapter 7 Feedback Unit 5

1. $\frac{5}{100}, \frac{12}{100}$
2. $\frac{35}{40}, \frac{12}{40}$
3. $\frac{12}{20}, \frac{45}{20}$
4. $\frac{15}{24}, \frac{-14}{24}$
5. $\frac{-21}{70}, \frac{40}{70}$
6. $\frac{-9}{24}, \frac{17}{24}$

Chapter 7 Feedback Unit 6

1. $\frac{39}{28}$
2. $\frac{-5}{4}$
3. $\frac{17}{18}$
4. $\frac{34}{7}$
5. $\frac{5}{9}$
6. $\frac{-5}{7}$
7. $\frac{8}{3}$
8. $\frac{39}{50}$
9. $\frac{23}{40}$

Chapter 7 Feedback Unit 7

1. a. $\frac{7}{2}$
   b. $\frac{1}{4}$
   c. $\frac{-4}{9}$
   d. $\frac{-1}{7}$
2. $\frac{6}{35}$
3. $\frac{-4}{15}$
4. $\frac{1}{7}$
5. $\frac{40}{3}$
6. $\frac{-15}{4}$
7. $\frac{15}{8}$

Chapter 7 Feedback Unit 8

1. $1.12
2. $262.50
3. $7,200
4. $6.30
5. $33.00

Chapter 7 Mastery Test

1. $\frac{-6}{0}$
2. $\frac{4}{13}$
3. $\frac{-3}{4}$
4. $\frac{-3}{14}$
5. 3
6. 42
7. $\frac{-8}{27}$
8. $\frac{-30}{7}$
9. $\frac{-2}{3}$
10. 2
11. -1
12. -157
13. $\frac{-1}{13}$
14. $\frac{7}{24}$
15. $\frac{-11}{36}$
16. $\frac{29}{72}$
17. $\frac{2}{245}$
18. $\frac{-27}{7}$
19. $\frac{16}{9}$
20. $\frac{27}{16}$
21. $18.90
22. $19.20

Chapter 8 Objectives

1. $11x - 23$
2. 18
3. $11x - 55$
4. $9x - 23$
5. $-2x - 8$

6. $-x + \frac{7}{9}$
7. x^{10}
8. x^5
9. x^{15}
10. $\frac{1}{x^5}$
11. x^4
12. x
13. $-64x^6y^3$
14. $125x^3y^{12}$
15. $-20x^4y^6$
16. $\frac{-2}{x}$
17. -1
18. $\frac{1}{5}$

Chapter 8 Feedback Unit 1

1. $x + \frac{1}{24}$
2. $x + \frac{1}{6}$
3. x
4. $x + \frac{5}{6}$
5. $x + \frac{5}{12}$
6. r
7. $\frac{4}{5}x + \frac{2}{21}$
8. $\frac{2}{3}x + \frac{13}{30}$

Chapter 8 Feedback Unit 2

1. $\frac{1}{2}x$
2. $\frac{-32}{15}x$
3. $\frac{-7}{12}x$
4. $\frac{14}{27}x$
5. $\frac{7}{10}x$
6. $\frac{-2}{3}x$
7. a
8. w
9. $3x$
10. $5x$

Chapter 8 Feedback Unit 3

1. yes
2. $3x + 28$
3. $\frac{1}{6}x - \frac{4}{3}$
4. $11x$
5. $5a$
6. $20x + 35$
7. $-21a - 12$
8. $5x - 10$
9. $-3x + 11$
10. $11x + 21$
11. $\frac{19}{4}x + \frac{23}{5}$
12. $5x - 4$

Chapter 8 Feedback Unit 4

1. x^{11}
2. x^9
3. x^{16}
4. $-6x^6$
5. $-5x^7$
6. $-15x^5y^2$
7. $-24x^9y^5$
8. $8x^9y^3$
9. $81x^8y^4$
10. $4x^2y^6$
11. $-625x^8y^4$
12. $625x^8y^4$
13. $-8x^3$
14. $-8x^3$

Chapter 8 Feedback Unit 5

1. x^3
2. x^3
3. $\frac{1}{x^4}$
4. 1
5. $\frac{1}{x^5}$
6. x^4
7. $\frac{1}{x^4}$
8. $\frac{x^2}{4}$
9. $-x$
10. $\frac{-3x^3}{2y^6}$
11. $\frac{x}{2z^4}$

Chapter 8 Feedback Unit 6

1. 23 inches
2. 26 meters
3. 85 sq yards
4. 18 sq centimeters
5. 240 cubic inches

Chapter 8 Mastery Test

1. $5x - 11$
2. $-17x + 37$
3. $2x - 14$

4. 4
5. $\frac{7}{2}x - \frac{11}{6}$
6. $\frac{59}{14}x - \frac{69}{8}$
7. x^8
8. x^8
9. x^{18}
10. $\frac{1}{x^3}$
11. x^4
12. $\frac{1}{x}$
13. $-56x^{12}$
14. $-3x^6$
15. $-16x^8y^4$
16. $27x^9y^6$
17. $-24x^5y^{10}$
18. $\frac{x^3}{2y}$
19. -1
20. $\frac{x}{y}$
21. 200 feet
22. 30 sq meters

CHAPTER 9
OBJECTIVES

1. $\frac{1}{12}$
2. $\frac{43}{40}$
3. $\frac{14}{15}$
4. $\frac{-7}{6}$
5. $\frac{11}{4}$
6. $\frac{-11}{5}$
7. $\frac{-19}{2}$
8. 16
9. -7
10. $\frac{15}{2}$
11. $\frac{15}{49}$
12. 9
13. $\frac{7}{5}$
14. $\frac{-13}{12}$
15. $\frac{10}{3}$
16. $\frac{-11}{4}$
17. $\frac{-4}{3}$
18. $\frac{21}{4}$; $\frac{5}{6} = \frac{5}{6}$ is true
19. $\frac{-1}{15}$; $\frac{7}{15} = \frac{7}{15}$ is true
20. 4; $\frac{17}{6} = \frac{17}{6}$ is true

CHAPTER 9
FEEDBACK UNIT 1

1. -4
2. -6
3. -1
4. 4
5. $\frac{19}{12}$
6. $\frac{9}{56}$
7. $\frac{9}{4}$
8. $\frac{31}{9}$

CHAPTER 9
FEEDBACK UNIT 2

1. a
2. b
3. $\frac{13}{5}$
4. $\frac{-4}{3}$
5. $\frac{5}{6}$
6. $\frac{7}{8}$
7. $\frac{-14}{3}$

CHAPTER 9
FEEDBACK UNIT 3

1. $\{\frac{1}{20}\}$
2. $\{\frac{11}{40}\}$
3. $\{\frac{-15}{28}\}$
4. $\{\frac{-20}{21}\}$
5. $\{\frac{-14}{5}\}$
6. $\{\frac{2}{3}\}$
7. $\{\frac{21}{5}\}$
8. $\{\frac{-4}{9}\}$
9. $\{\frac{9}{7}\}$
10. {-1}
11. {-11}

CHAPTER 9
FEEDBACK UNIT 4

1. $\frac{-41}{18}$; $\frac{5}{6} = \frac{5}{6}$ is true
2. $\frac{23}{8}$; $\frac{3}{10} = \frac{3}{10}$ is true
3. -2; $\frac{-5}{3} = \frac{-5}{3}$ is true
4. $\frac{2}{3}$; $\frac{-3}{8} = \frac{-3}{8}$ is true
5. $\frac{1}{4}$; $\frac{1}{10} = \frac{1}{10}$ is true
6. $\frac{-1}{7}$; $\frac{-3}{8} = \frac{-3}{8}$ is true
7. $\frac{-3}{5}$; $\frac{23}{60} = \frac{23}{60}$ is true
8. $\frac{1}{3}$; $\frac{-8}{21} = \frac{-8}{21}$ is true

Chapter 9
Feedback Unit 5

1. 37
2. 32
3. 8
4. 4
5. 7

Chapter 9
Mastery Test

1. $\frac{-1}{5}$
2. $\frac{23}{4}$
3. -4
4. -12
5. $\frac{61}{2}$
6. $\frac{-13}{5}$
7. $\frac{2}{3}$
8. $\frac{1}{6}$
9. $\frac{7}{8}$
10. $\frac{-3}{7}$
11. $\frac{-2}{25}$
12. $\frac{-5}{7}$
13. $\frac{-3}{2}$
14. $\frac{5}{7}$
15. $\frac{-29}{16}$
16. $\frac{2}{5}$
17. $\frac{7}{6}$
18. $\frac{1}{3}$; $\frac{-3}{10} = \frac{-3}{10}$ is true
19. $\frac{-5}{23}$; $\frac{-37}{92} = \frac{-37}{92}$ is true
20. $\frac{6}{7}$; $\frac{26}{35} = \frac{26}{35}$ is true
21. 8
22. 4

Index

A

Addends 78
Adding with a common denominator 285
Addition 9
Addition of integers 141-150
Associative Law of Addition 80, 190, 306
Associative Law of
  Multiplication 88, 195, 308

B

Base 168, 208, 321
Between 2
Both sides of an equation 227
Braces 2

C

Cancelling 269
Changing names of rational numbers 276
Checking a
  solution 122-125, 239-240, 366-368
Coefficient 210, 313, 324, 351
Common denominators 279-281
Commutative Law of Addition 78, 190, 305
Commutative Law of
  Multiplication 86, 194, 308
Complex fractions 294
Composite numbers 175
Counting numbers 6
Cubed 168

D

Denominator 253
Difference 19
Distributive Law of Multiplication
  over Addition 92, 200, 312
Dividing power expressions 332
Division 34
Division of rationals 293

E

Element 4
Empty set 3, 118, 238
Equal rational numbers 256
Equation 111, 221
Equivalent 190
Equivalent equations 226, 344, 348
Equivalent expressions 77, 78
Evaluate
  a division expression 36
  a multiplication expression 31
  a subtraction expression 23
  an addition expression 10
Evaluating a multiplication 266
Evaluating numerical expressions 41
Evaluating open expressions 72, 185-188
Evaluation 185
Even numbers 2
Exponent 168. 208, 313
Expression 9

F

Factors 50, 51, 167, 261, 320
Fourth power 208

H

Highest common factor (HCF) 52, 277

I

Improper fraction 248
Integers 135
Integers as rational numbers 255

L

Least common multiple (LCM) 55, 278, 362
Left side of an equation 225
Left side of the equals sign 114
Like terms 93, 192

M

Mathematical statement 110, 189
Member 4
Minus sign 158
Mixed number 248
Multiple 54
Multiplication 27
Multiplication expression 27
Multiplication Law of Negative One 199, 311
Multiplication Law of One 90, 198, 310
Multiplication of integers 151-157
Multiplication symbol (•) 29, 71

N

Negative 134
Negative integers 135
Number line 139
Numerator 252
Numerical expression 40, 65, 185

O

Odd numbers 3
Open expression 66, 185
Open sentence 110, 189
Opposite 134, 286, 344
Order of Operations 44, 171

P

Pair of factors 173
Parentheses 10, 22
Positive integer factors 173
Positive integers 135
Power expression 209, 321
Prime number 174
Product 28

Q

Quotient 37

R

Raising a power to a power 328
Ratio 221
Rational numbers 221
Reciprocals 291, 348
Removing parentheses 97, 204
Replacement set 116

S

Second power 168, 208
Set 1
Set of integers 135
Simplest name for a rational number 262
Simplify 191-193, 196, 201-203, 206-207
Simplifying Addition Expressions 306-307
Simplifying Multiplication Expressions 309-311
Simplifying open expressions 83, 88, 92, 94, 99
Simplifying power multiplications 212-213
Simplifying rational numbers 263-265
Solution 231
Solution of an equation 114
Solve 346
Solving equations 111, 231, 343-366
Square brackets 13
Squared 168, 208
Statement 221
Substitution 79
Subtract 18
Subtraction of integers 158-166
Sum 9

T

Third power 168, 208
Total 26
Truth set 116, 238, 356

V

Variable 71